The Long-Term Wilderness Survival Bible

10 Books in 1

The Complete Collection to Survive in Harsh Environments | Techniques for Building Shelter, Securing Water, and Sustain Life Far from Civilization

Joseph Scott

Copyright © 2024

All rights reserved. No part of this guide may be reproduced in any form without permission in writing from the publisher, except for brief quotations used for publishable articles or reviews.

Legal Disclaimer

The information contained in this book and its contents is not designed to replace any form of medical or professional advice; and is not meant to replace the need for independent medical, financial, legal, or other professional advice or services that may be required. The content and information in this book have been provided for educational and entertainment purposes only.

The content and information contained in this book have been compiled from sources deemed reliable, and they are accurate to the best of the Author's knowledge, information, and belief. However, the author cannot guarantee its accuracy and validity and therefore cannot be held liable for any errors and/or omissions. Further, changes are periodically made to this book as needed. Where appropriate and/or necessary, you must consult a professional (including but not limited to your doctor, attorney, financial advisor, or other such professional) before using any of the suggested remedies, techniques, and/or information in this book.

Upon using this book's contents and information, you agree to hold harmless the Author from any damages, costs, and expenses, including any legal fees, potentially resulting from the application of any of the information in this book. This disclaimer applies to any loss, damages, or injury caused by the use and application of this book's contents, whether directly or indirectly, whether for breach of contract, tort, negligence, personal injury, criminal intent, or under any other circumstance.

You agree to accept all risks of using the information presented in this book.

You agree that by continuing to read this book, where appropriate and/or necessary, you shall consult a professional (including but not limited to your doctor, attorney, financial advisor, or other such

Table of Contents

Introduction..1

Book 1:
Survival Shelter Construction...3
Chapter 1: Understanding Shelter Design Fundamentals..3
Chapter 2: Utilizing Natural Materials for Shelter...7
Chapter 3: Building Techniques for Long-Term Shelter..10
Chapter 4: Insulating and Weatherproofing Your Refuge...14

Book 2:
Water Purification Essentials..17
Chapter 1: Finding and Identifying Safe Water Sources...17
Chapter 2: Traditional Water Purification with Boiling and Distillation.......................20
Chapter 3: Chemical Water Treatment with Iodine and Chlorine................................23
Chapter 4: DIY and Natural Filtration Systems...26

Book 3:
Foraging for Survival..29
Chapter 1: The Forager's Guide to Plant Identification..29
Chapter 2: Regional Edible Plants and Their Discovery...32
Chapter 3: Avoiding Toxic Plants and Dangerous Lookalikes......................................35

Book 4:
Hunting and Trapping Skills..38
Chapter 1: Beginner's Guide to Basic Hunting Techniques...38
Chapter 2: Constructing Effective Traps for Game...42
Chapter 3: Mastering the Art of Tracking and Stalking..45
Chapter 4: Respecting Wildlife with Ethical Hunting Practices...................................48

Book 5:
Fire Making Mastery...51
Chapter 1: The Science Behind Fire Basics..51
Chapter 2: Modern Fire Starters for Quick Lighting...54
Chapter 3: Maintaining and Safeguarding Your Fire..57
Chapter 4: Creating Emergency Signal Fires for Rescue..60

Book 6:
Survival Navigation..63
Chapter 1: Map-Free Navigation Tips and Tricks..63
Chapter 2: Crafting Improvised Compasses for Direction...66
Chapter 3: Safe and Effective Night Navigation Strategies..69

Book 7:
Wilderness First Aid...73

Chapter 1: Essential Elements of a Survival First Aid Kit...73
Chapter 2: Dealing with Common Injuries in the Wild..77
Chapter 3: Natural Remedies Using Medicinal Plants..80
Chapter 4: Managing Environmental Hazards Like Hypothermia and Heatstroke............................83

Book 8:
Weather and Environment Adaptation..86
Chapter 1: Predicting Weather Changes and Preparing Accordingly...86
Chapter 2: Survival Strategies for Extreme Cold Conditions...90
Chapter 3: Life-Saving Techniques for Arid Climate Survival...93
Chapter 4: Adapting to Wetland and Rainforest Environments..96

Book 9:
Long-Term Food Preservation..99
Chapter 1: Traditional Meat Preservation with Smoking and Curing...99
Chapter 2: Drying and Dehydrating a Variety of Foods... 102
Chapter 3: Fermenting and Pickling for Longevity.. 105
Chapter 4: Natural Underground Storage Techniques.. 108

Book 10:
Living Off the Grid.. 111
Chapter 1: Fundamentals of Sustainable Off-Grid Living.. 111
Chapter 2: Harnessing Renewable Energy in the Wilderness.. 115
Chapter 3: Growing Your Own Food with Sustainable Agriculture... 118
Chapter 4: Effective Waste Management and Recycling Practices.. 121

Introduction

In the vast expanse of the wilderness, where nature unfolds in both its majestic beauty and raw ferocity, lies the ultimate test of human resilience, adaptability, and survival instincts. "LONG TERM WILDERNESS SURVIVAL BIBLE" is not just a book; it's a comprehensive guide that embarks you on a profound journey into mastering the art of thriving in the wild for extended periods. This manual is crafted for those who seek to embrace the wilderness not as a brief escape but as a domain where they can coexist with nature, relying on their skills and knowledge to sustain life.

In an era where the disconnect between humans and the natural world grows wider, a phenomenon has emerged—a resurgent desire to reconnect with the earth, to understand its rhythms, and to survive within its embrace. This book responds to that call, providing a detailed roadmap to becoming self-sufficient in the wilderness, far from the conveniences of modern civilization. Whether you're an avid outdoorsman, an adventurer seeking to explore the depths of the wild, or someone who wishes to prepare for any scenario, this book is an indispensable companion.

From the construction of sturdy, weather-resistant shelters that offer refuge, to the essentials of purifying water to sustain life; from the art of foraging for edible plants to the tactics of hunting and trapping for food; this book covers every critical aspect of long-term wilderness survival. It delves into the mastery of fire-making, a skill as ancient as humanity itself, and navigates through the intricacies of survival navigation, ensuring you never lose your way. Wilderness first aid is explored in depth, equipping you with the knowledge to address injuries and health emergencies with confidence.

Moreover, this survival bible offers invaluable insights into adapting to various weather and environmental conditions, ensuring you can withstand the challenges posed by changing climates and landscapes. It guides you through the practices of long-term food preservation, enabling you to store and sustain your food supplies over time. Lastly, it ventures into the realm of living off the grid, teaching you how to establish a sustainable existence in harmony with the natural world.

By immersing yourself in this book, you will learn:

- Survival Shelter Construction: The principles of designing and building shelters that withstand the elements, offering safety and comfort.
- Water Purification Essentials: Techniques for securing safe drinking water, a cornerstone of survival in the wilderness.
- Foraging for Survival: How to identify and harvest edible plants, nuts, and fruits, turning the wilderness into a bountiful pantry.

- Hunting and Trapping Skills: Strategies for tracking, hunting, and trapping wildlife, ensuring a reliable food source.
- Fire Making Mastery: Mastering the ancient skill of fire-making, crucial for warmth, cooking, and protection.
- Survival Navigation: Navigating the wilderness with or without a compass, using the sun, stars, and natural landmarks to guide your way.
- Wilderness First Aid: Managing medical emergencies and providing first aid with limited resources.
- Weather and Environment Adaptation: Adapting to various climates and environments, from scorching deserts to frozen tundras.
- Long-Term Food Preservation: Preserving food through drying, smoking, fermenting, and other methods to ensure sustainability.
- Living Off the Grid: Establishing a self-sufficient, sustainable lifestyle in harmony with nature.

"LONG TERM WILDERNESS SURVIVAL BIBLE" is more than a survival manual; it's a testament to the human spirit's capacity to endure, adapt, and thrive in the face of the raw forces of nature. It empowers you with the skills, knowledge, and mindset required to face the wilderness not with fear, but with confidence and respect. This book is a call to rediscover the primal connection to the earth, to learn from it, and ultimately, to survive within it.

Book 1:
Survival Shelter Construction

Chapter 1: Understanding Shelter Design Fundamentals

In the realm of survival, shelter stands as the guardian against the harsh elements of nature. It is not merely a structure but a sanctuary that preserves warmth, offers protection, and provides psychological comfort. Understanding the fundamentals of shelter design is paramount, as it lays the groundwork for constructing a refuge that could mean the difference between life and peril. This chapter delves into the essential principles of shelter design, focusing on the critical aspects that ensure a shelter's effectiveness and sustainability.

The Purpose of a Shelter

Before diving into the technicalities, it's essential to understand a shelter's primary purposes:
- Protection from the Elements: A shelter's foremost function is to protect from wind, rain, snow, and sun.
- Conservation of Heat: Retaining body heat is vital in cold environments, making insulation a key factor in design.
- Safety from Predators: A shelter should offer safety from wildlife, providing a secure place to rest.
- Psychological Comfort: Beyond physical needs, a shelter provides a sense of safety and normalcy in survival situations.

Location Selection

Choosing the right location is the first critical step in shelter design. Consider the following factors:
- Safety: Avoid areas prone to natural hazards such as flooding, avalanches, or falling rocks.
- Water Availability: Proximity to a water source is crucial, but don't build too close to avoid flooding risks.
- Wind Exposure: Seek locations that are sheltered from prevailing winds to reduce thermal loss.
- Sun Exposure: In colder climates, a sunny spot can help warm the shelter during the day.

Design Principles

The design of a survival shelter is influenced by several key principles:
- Simplicity: A shelter should be as simple as possible to conserve energy and resources.

- Efficiency: The design should maximize warmth and protection with minimal materials and effort.
- Durability: Even temporary shelters should be sturdy enough to withstand the elements during the intended use period.
- Insulation: Proper insulation is essential to retain heat and maintain a comfortable internal temperature.

Types of Shelters

There are various shelters suited to different environments and situations. Some common types include:
- Lean-to Shelter: Simple and quick to build, offering protection from wind and rain.
- A-Frame Shelter: Provides better insulation and protection from the elements on all sides.
- Snow Cave: An effective option in snowy environments, offering excellent insulation.
- Debris Hut: Utilizes natural materials for insulation, suitable for forested areas.

Materials and Tools

The availability of materials plays a significant role in the type of shelter you can construct. Commonly used natural materials include:
- Wood: For framework and support.
- Leaves and Grass: For insulation and waterproofing.
- Snow: For insulation in winter shelters.

Basic tools might include:
- Knife: For cutting materials.
- Rope or Vines: For securing structures.
- Shovel: For digging and snow shelters.

Construction Techniques

Effective shelter construction relies on proven techniques:
- Foundation: Start with a solid foundation to ensure stability.
- Framework: Construct a sturdy frame using branches or other materials.
- Covering: Use leaves, branches, or snow to cover the frame, focusing on thick layers for insulation.
- Entrance: Design the entrance to minimize heat loss, using a small size or creating a windbreak.

Insulation and Weatherproofing

Insulation is key to a shelter's effectiveness, particularly in cold environments. Techniques include:
- Layering: Use multiple layers of materials to trap air and provide insulation.
- Packing Snow: In snow shelters, packing the snow can increase its insulating properties.
- Sealing Gaps: Ensure the shelter is tightly sealed to prevent wind and water ingress.

Ventilation

While insulation is crucial, proper ventilation is also necessary to prevent condensation and ensure a supply of fresh air. Strategically placed vents or an open entrance can facilitate airflow without significantly compromising warmth.

Environmental Impact

Building with respect to the environment is essential. Principles include:
- Use Renewable Resources: Opt for materials that are abundant and renewable.
- Minimize Disturbance: Alter the environment as little as possible, preserving the natural habitat.
- Leave No Trace: Dismantle the shelter and scatter materials to minimize your impact.

Psychological Considerations

The psychological impact of a well-designed shelter cannot be overstated. It provides a sense of security and control, boosting morale in survival situations. The process of building and inhabiting the shelter can also offer a mental distraction from the stresses of survival.

Practical Applications

Understanding these fundamentals is just the beginning. The real learning comes from applying this knowledge in the field. Practice building different types of shelters in various environments to gain experience and confidence. Remember, the best shelter is one that meets your immediate needs while conserving your energy and resources for the challenges ahead.

In conclusion, shelter is a fundamental human need, especially in survival situations. This chapter has laid the groundwork for understanding shelter design fundamentals, emphasizing the importance of location, efficiency, and adaptability. As we progress through this book, we'll build on these principles, exploring more advanced techniques and strategies for constructing shelters that offer not just survival, but a semblance of comfort in the wilderness. Remember, the skills you develop through practice can one day turn a dire situation into a manageable one, underlining the importance of preparation, knowledge, and respect for nature's power.

Chapter 2: Utilizing Natural Materials for Shelter

The essence of survival often lies in our ability to adapt and utilize what nature offers. When constructing a shelter, the materials provided by the surrounding environment are not only resources but lifelines. This chapter delves into the art of harnessing natural materials for shelter construction, highlighting the importance of sustainable practices and resourcefulness. By understanding how to effectively use what the land offers, you can create shelters that are not only functional but also harmonious with the environment.

The Philosophy of Natural Materials

Utilizing natural materials for shelter construction is rooted in the principles of sustainability and efficiency. This approach requires a deep understanding of the environment and a respectful interaction with nature. The philosophy is simple: use what is abundant, renewable, and least impactful to the environment, ensuring that your survival practices contribute to the preservation of natural habitats.

Identifying and Collecting Materials

The first step in utilizing natural materials is to identify and collect what is available in your surroundings. This process varies significantly depending on the ecosystem you are in:

- Forests and Woodlands: Look for branches, leaves, moss, and bark. Fallen trees can provide both structure and insulation.
- Grasslands: Grasses, reeds, and sod are abundant and can be used for roofing and insulation.
- Deserts: While materials are scarcer, you can use sand, rocks, and sparse vegetation like cacti for constructing sun shelters.
- Snowy Environments: Snow is an excellent insulator and can be used to build shelters like igloos and quinzhees.

Working with Wood

Wood is one of the most versatile materials for shelter construction, offering both structural support and insulation. Here are some tips for working with wood:
- Selecting Wood: Look for strong, dry branches for the frame. Green wood, although flexible, may not be as sturdy.

- Tools: While natural shelters often don't require tools, having a knife or hatchet can make working with wood easier.
- Lashing and Tying: Learn basic knots and lashing techniques to secure branches together without nails or modern fasteners.

Using Leaves, Grass, and Moss for Insulation

Leaves, grass, and moss can serve as excellent insulating materials, trapping air and retaining heat. Here's how to use them effectively:
- Thick Layers: Pile these materials thickly on the shelter's exterior to maximize insulation.
- Waterproofing: When layered correctly, leaves and grass can also help waterproof your shelter, directing water away from the interior.
- Renewable Resources: Always collect these materials in a way that allows the plant life to regenerate.

Constructing with Snow

In winter environments, snow can be both a challenge and a resource. Here's how to use snow to your advantage:
- Insulation: Snow shelters, like igloos and snow caves, use the insulating properties of compacted snow to maintain warmth.
- Construction Techniques: Learning to cut and stack snow blocks or hollow out a snow cave can be lifesaving in cold climates.

Incorporating Rocks and Earth

Rocks and earth can provide excellent windbreaks and structural support for shelters. Here are some considerations:
- Stability: Use larger rocks for the foundation to stabilize your shelter.
- Thermal Mass: Rocks can absorb heat during the day and release it at night, helping to regulate temperature.

Advantages of Natural Materials

- Sustainability: Using materials that are readily available reduces the need to carry or find manufactured materials.
- Cost-Effective: Natural materials are free and can be found in abundance with a little knowledge and effort.
- Environmental Impact: Building with natural materials minimizes your environmental footprint, especially when following principles of leave-no-trace ethics.

Challenges and Considerations

While natural materials offer numerous benefits, there are also challenges:
- Time and Energy: Gathering and preparing materials can be time-consuming and physically demanding.
- Skill and Knowledge: Knowing which materials to use and how to use them effectively requires practice and understanding of the local environment.
- Durability: Natural shelters may not be as long-lasting as those built with manufactured materials and may require maintenance.

Practical Tips

- Practice Makes Perfect: Regularly practice building shelters in different environments to hone your skills.
- Be Creative: Sometimes, the best solutions come from thinking outside the box and using materials in unconventional ways.
- Respect Nature: Always gather materials sustainably, taking only what you need and leaving the environment as undisturbed as possible.

Utilizing natural materials for shelter is a testament to human ingenuity and our deep connection to the earth. This chapter has provided a foundation for understanding how to effectively and sustainably use what the environment offers to protect ourselves. As we move through the wilderness of our world, let us do so with respect, gratitude, and the knowledge that nature is not just a resource but a partner in our survival. Remember, the skills and practices outlined here are not just for emergency situations but also for enriching our outdoor experiences and fostering a sustainable relationship with the natural world.

Chapter 3: Building Techniques for Long-Term Shelter

Building a long-term shelter is a testament to human resilience and ingenuity. It's about creating a sustainable living space that can withstand the elements and provide comfort over an extended period. This chapter explores the various techniques and considerations for constructing long-term shelters, emphasizing the importance of planning, durability, and adaptability to the environment. Whether you're in a survival situation or simply seeking to connect with nature on a deeper level, these insights will guide you in establishing a robust and efficient refuge.

Planning and Design

Before embarking on the construction of a long-term shelter, thorough planning is crucial. Consider the following:

- Site Selection: Choose a location based on safety, water availability, exposure to sunlight, and protection from the elements.
- Materials: Decide on materials based on what's available in the surrounding environment and the durability required for long-term use.
- Design: Tailor the design to your specific needs, climate, and terrain. Consider space for sleeping, cooking, and storage.

Foundations

A solid foundation is critical for the longevity and stability of any long-term shelter. Techniques include:

- Elevated Platforms: In damp or flood-prone areas, building an elevated platform can keep the living area dry.
- Stone Foundations: Stones can create a sturdy base that prevents wood rot and deters pests.

Framing Techniques

The frame of your shelter provides structure and support. Several techniques can be used, depending on the materials available:

- Post-and-Beam Construction: This technique involves setting large posts into the ground and connecting them with horizontal beams, suitable for larger structures.

- A-Frame: An A-frame structure is sturdy and effective at shedding snow, making it ideal for snowy climates.

Roofing

A well-constructed roof protects from rain, sun, and snow. Options include:

- Pitched Roofs: Sloped roofs are effective at draining rainwater and snow.
- Thatched Roofing: Using grass, reeds, or palm leaves can create a waterproof and insulating roof, ideal for certain climates.

Walls

The walls of your shelter not only provide protection from the elements but also insulation. Techniques include:

- Wattle and Daub: A framework of woven branches (wattle) covered with a mixture of mud, straw, and dung (daub) can create sturdy, insulated walls.
- Cob: Similar to wattle and daub, cob is a mixture of clay, sand, straw, and water, shaped into walls without a wooden framework.

Insulation and Weatherproofing

Effective insulation is key to maintaining a comfortable interior temperature. Weatherproofing protects against water and wind:

- Natural Insulation: Use materials like moss, leaves, and grass for insulating walls and roofs.
- Mud Plaster: A coating of mud plaster can seal walls, providing additional insulation and waterproofing.

Doors and Windows

Properly designed doors and windows regulate temperature, light, and air flow:

- Orientation: Position windows for maximum light and warmth, especially in colder climates.

- Size and Coverings: Small windows reduce heat loss. Shutters or animal hides can cover openings when needed.

Flooring

A good floor keeps the interior dry and comfortable. Options include:

- Raised Floors: Elevated wooden floors protect against dampness and pests.
- Earth Floors: Compacted earth, especially when treated with linseed oil, can create a durable and natural floor.

Sustainability and Environmental Impact

Building a long-term shelter is an opportunity to practice sustainability:

- Renewable Resources: Use materials that can be replenished or that minimize environmental impact.
- Eco-Friendly Techniques: Employ construction methods that do not harm the environment, such as using hand tools and avoiding chemicals.

Maintenance and Upkeep

Long-term shelters require regular maintenance:

- Inspections: Regularly check the structure for signs of wear, damage, or pest infestation.
- Repairs: Promptly address any issues to prevent minor problems from becoming major ones.

Adapting to the Environment

Your shelter should be a living part of the environment, adapting to changes and utilizing natural benefits:

- Natural Cooling and Heating: Use landscaping, such as trees for shade or positioning the shelter to catch prevailing breezes.
- Water Collection: Design the shelter with rainwater collection in mind, using gutters and storage containers.

Personalization and Comfort

A long-term shelter is more than a structure; it's a home:

- Interior Design: Create a space that reflects your needs and personality, using natural materials for furnishings and decorations.
- Outdoor Spaces: Consider adding outdoor living areas, such as a fire pit or a covered seating area.

Building a long-term shelter is a profound connection to our ancestral past and a practical expression of our desire to live in harmony with nature. This chapter has provided a framework for constructing shelters that are not just about survival but about creating a sustainable and comfortable living space. By applying these techniques with respect for the environment and a spirit of innovation, you can build a shelter that stands as a testament to human creativity and resilience. Remember, the ultimate goal is to create a space that meets your needs while respecting the delicate balance of the natural world around you.

Chapter 4: Insulating and Weatherproofing Your Refuge

In the construction of any long-term shelter, insulating and weatherproofing are critical components that transform a basic structure into a sustainable and comfortable living space. These processes are essential for protecting against the harshness of the elements, conserving energy, and ensuring the durability of your shelter. This chapter dives into the methods and materials that can be utilized to insulate and weatherproof your refuge effectively, ensuring it remains a haven of warmth and safety regardless of the external conditions.

Understanding Insulation and Weatherproofing

- Insulation: Insulation involves creating a barrier of material that traps air, slowing the transfer of heat and helping to maintain a consistent internal temperature.
- Weatherproofing: Weatherproofing protects against water, wind, and other environmental elements that could compromise the integrity and comfort of your shelter.

Principles of Effective Insulation

- Thermal Resistance: The ability of an insulating material to resist heat flow, typically measured in R-values—the higher the R-value, the better the insulation.
- Comprehensive Coverage: Ensuring that walls, roof, and floor are all adequately insulated to prevent thermal bridges and heat loss.
- Material Selection: Choosing insulation materials that are suitable for the climate and the available resources.

Natural Insulating Materials

- Leaves and Moss: Great for filling gaps and covering shelters, providing a dense layer of insulation.
- Grass and Straw: Used in thatched roofs and walls, these materials trap air and create an effective insulating barrier.
- Animal Hides: Can be used to cover openings, adding warmth and reducing airflow.
- Earth and Mud: When applied to walls, they provide mass insulation, absorbing heat during the day and releasing it at night.

Techniques for Applying Insulation

- Layering: Building up layers of natural materials to increase insulation. This can be particularly effective in roofs and walls.
- Packing: Tightly packing materials like leaves, grass, or snow around and on top of the shelter to reduce air movement and heat loss.
- Creating Air Spaces: Constructing walls with cavities or air spaces that act as additional insulating barriers.

Weatherproofing Strategies

- Roofing: Ensuring a steep enough pitch for water runoff and using overlapping materials like bark or large leaves to direct water away from the shelter.
- Sealing Gaps: Filling cracks and gaps with mud, moss, or other pliable materials to prevent water ingress and reduce wind penetration.
- Drainage: Establishing clear drainage paths around the shelter to prevent water accumulation that could lead to flooding or moisture problems.

Enhancing Weather Resistance

- Wind Barriers: Constructing windbreaks using natural terrain, vegetation, or additional structures to reduce wind impact on the shelter.
- Water Repellents: Applying natural oils or resins to materials to increase their water resistance.
- Regular Maintenance: Inspecting and maintaining the shelter regularly to address any issues before they become significant problems.

Sustainable Practices

- Using Locally Sourced Materials: Minimizes the environmental impact and ensures that the materials are well-suited to the local climate.
- Renewable Resources: Choosing materials that can be replenished or that have minimal impact on the environment.
- Energy Conservation: Designing the shelter to make the most of natural heating and cooling, reducing the need for additional energy inputs.

Ventilation and Moisture Control

- Balancing Insulation with Ventilation: Ensuring adequate ventilation to prevent condensation buildup, which can lead to mold and mildew.
- Moisture Barriers: Using materials that naturally repel water or constructing barriers to prevent moisture from seeping into the shelter.
- Air Circulation: Designing openings or vents that can be adjusted to control airflow and manage humidity levels.

Practical Applications

- Seasonal Adjustments: Modifying insulation and weatherproofing techniques based on seasonal changes, such as adding extra layers of insulation for winter.
- Adapting to Climate: Tailoring strategies to the specific challenges of the local climate, whether it's heat, cold, humidity, or dryness.
- Integration with Natural Surroundings: Using the landscape and natural features to enhance the shelter's insulation and weatherproofing, such as building into a hillside for natural insulation.

Troubleshooting Common Issues

- Leaks and Water Damage: Identifying and repairing leaks promptly to prevent structural damage and mold growth.
- Inadequate Insulation: Assessing and augmenting insulation if the shelter is losing too much heat or becoming too hot.
- Ventilation Problems: Adjusting ventilation strategies to address issues with air quality or moisture accumulation.

Insulating and weatherproofing your refuge are vital processes that require careful consideration and ongoing attention. By understanding and applying the principles and techniques outlined in this chapter, you can ensure that your shelter is not only a place of protection but also a comfortable and sustainable home. The key lies in choosing the right materials, employing effective construction techniques, and adapting to the ever-changing environment. Remember, a well-insulated and weatherproofed shelter not only provides physical comfort but also peace of mind, allowing you to thrive in harmony with nature..

Book 2:
Water Purification Essentials

Chapter 1: Finding and Identifying Safe Water Sources

In the quest for survival and sustainable living, one cannot overemphasize the importance of water. Not just any water, but clean, safe water is crucial for health and well-being. The first chapter of "Water Purification Essentials" begins with the fundamental skill of finding and identifying safe water sources. This knowledge is not only vital in survival situations but also for anyone exploring the outdoors or seeking to live more autonomously.

The Vital Role of Water

Water is the cornerstone of life. It hydrates, aids in digestion, regulates body temperature, and supports every cellular function in our bodies. However, accessing clean water can be a challenge in wilderness settings or during emergencies. Here's why understanding water sources is essential:

- Hydration: Prevents dehydration and its severe complications.
- Health: Reduces the risk of contracting waterborne diseases.
- Survival: Increases chances of survival in emergency situations.

Locating Water Sources

The ability to locate water sources is your first step towards ensuring a supply of potable water:

- Natural Water Bodies: Rivers, lakes, streams, and springs are primary sources. Moving water is generally preferable to stagnant water.
- Rainwater: Collecting rainwater can provide a clean source, especially when collected directly from the sky or through clean catchment systems.
- Dew and Condensation: Early morning or late evening are prime times for collecting dew from vegetation or creating condensation traps.
- Ground Moisture: In arid regions, digging in dry riverbeds or valleys can uncover underground moisture.

Identifying Safe Water Sources

Not all water is safe for consumption. Use these guidelines to assess water safety:

- Clarity: Clear water is preferred, though not a definitive indicator of safety.
- Smell: Avoid water with foul odors, indicating contamination.
- Surroundings: Water found in nature should be upstream from any potential pollution sources, such as agricultural fields or industrial areas.

Assessment Techniques

After locating a potential water source, it's crucial to assess its safety:

- Visual Inspection: Look for signs of algae blooms or floating debris, which can indicate poor water quality.
- Animal Life: The presence of healthy animal and insect activity can suggest water quality, but it's not foolproof.
- Vegetation: Lush vegetation often thrives near reliable water sources, but beware of plants that thrive in contaminated water.

Collection Methods

Efficient collection methods can enhance the safety of the water:

- Rainwater Harvesting: Use clean containers or tarps to collect rainwater, avoiding surfaces that may introduce contaminants.
- Dew Collection: Absorbent materials like cloths or sponges can gather dew from vegetation, which can then be wrung out.
- Water from Vegetation: Techniques like tying a bag around a leafy branch can collect transpiration water, although the yield is often low.

Understanding Risks

Awareness of the risks associated with unclean water is critical:

- Pathogens: Bacteria, viruses, and parasites found in untreated water can cause severe illness.
- Chemicals: Pesticides, heavy metals, and pollutants from industrial and agricultural runoff can contaminate water sources.
- Environmental Changes: Seasonal variations and human activities can alter the safety of natural water sources over time.

Safety Precautions

Taking safety precautions can mitigate risks:

- Avoid Stagnant Water: Stagnant water is more likely to harbor pathogens and pests.
- Use Natural Filters: Sand, gravel, and charcoal can provide rudimentary filtration to remove large particulates before purification.
- Boil or Treat Water: Always purify water through boiling, chemical treatment, or filtration before consumption, regardless of the source's apparent cleanliness.

The ability to find and identify safe water sources is an invaluable skill in our increasingly unpredictable world. Whether you are a hiker, camper, or someone interested in off-grid living, understanding the principles laid out in this chapter is the first step toward ensuring access to safe drinking water. As we move forward in this book, we will delve deeper into how to make these water sources safe for drinking through traditional boiling, distillation, chemical treatment, and innovative DIY filtration systems. This foundational knowledge sets the stage for a deeper exploration of water purification techniques that can safeguard your health and enhance your self-sufficiency in any situation.

Chapter 2: Traditional Water Purification with Boiling and Distillation

After finding and identifying a potential water source, the next critical step is purification. Traditional methods, notably boiling and distillation, have stood the test of time for their effectiveness in making water safe for consumption. This chapter delves into these age-old techniques, offering a practical guide to purifying water in various circumstances.

Boiling: Nature's Purifier

Boiling is one of the simplest and most reliable methods to purify water. It kills pathogens such as bacteria, viruses, and parasites, making the water safe to drink.

- How to Boil Water for Purification:
  - Bring water to a rolling boil for at least one minute at low altitudes and three minutes at altitudes above 2,000 meters (6,562 feet), where water boils at a lower temperature.
  - Let the water cool naturally before consumption to avoid burns.
  - Consider covering the pot to reduce evaporation and conserve fuel.

- Advantages of Boiling:
  - Effectiveness: Kills most types of pathogens.
  - Accessibility: Requires only a heat source and a container.
  - Cost: Minimal, especially if using natural fuels.

- Limitations:
  - Does not remove chemical pollutants or heavy metals.
  - Requires a significant amount of fuel and time, especially in large quantities.
  - Leaves the taste of the water unchanged, which may be undesirable in some cases.

Distillation: The Ultimate Purifier

Distillation goes a step further than boiling by vaporizing water and then condensing it back into liquid form. This process not only kills pathogens but also removes chemicals, salts, and heavy metals, making it one of the most comprehensive purification methods.

- How to Distill Water:
  - Boil water and direct the steam into a cooling tube or surface. As the steam cools, it condenses back into liquid water, which is collected in a clean container.
  - DIY distillation setups can be made with pots, pans, and tubing, or by using solar distillation techniques.
  - Solar distillation utilizes the sun's heat to evaporate water, which then condenses on a cool surface, such as plastic sheeting, and drips into a collection container.

- Advantages of Distillation:
  - Removes a wide range of contaminants, including pathogens, chemicals, and salts.
  - Provides the purest form of water, which is especially important in areas with heavy contamination.

- Limitations:
  - Requires more equipment and technical knowledge than boiling.
  - Energy-intensive, needing a constant heat source for evaporation.
  - Time-consuming, particularly for large quantities of water.

Practical Applications and Tips

Both boiling and distillation are invaluable in different scenarios, from emergency situations to everyday use in off-grid living. Here are some practical tips for implementing these methods:

- Fuel Efficiency: When boiling water, use a lid to reduce evaporation and save fuel. For distillation, consider solar options to minimize fuel use.
- Improving Taste: Boiled water can taste flat due to the lack of dissolved oxygen. Shaking it or pouring it back and forth between two containers can help improve the taste.
- Use of Containers: Ensure all containers used in boiling and distillation are clean and free from chemical contaminants.
- Solar Distillation: Can be an effective method in sunny climates, requiring minimal equipment and no fuel.

Considerations for Use

When deciding between boiling and distillation, consider the specific needs and available resources:

- Water Quality: If chemical contamination is a concern, distillation may be the better option.

- Fuel Availability: Boiling is more feasible where fuel is readily available; in fuel-scarce regions, solar distillation might be more practical.
- Quantity Needed: For small quantities, boiling is quick and efficient. For purifying large volumes or when needing the highest purity level, distillation, though more time-consuming, is superior.

Understanding and applying the principles of boiling and distillation can empower individuals to ensure their water supply is safe, enhancing self-sufficiency and preparedness. These traditional methods, grounded in the basic principles of heat and physics, offer a line of defense against waterborne diseases and contaminants. As we advance through "Water Purification Essentials," we'll explore modern and innovative approaches to complement these time-honored techniques, ensuring readers have a comprehensive toolkit for accessing clean water in any situation.

Chapter 3: Chemical Water Treatment with Iodine and Chlorine

In the wilderness or during emergencies, chemical treatment is a practical and effective method for purifying water. This chapter explores the use of iodine and chlorine, two chemicals widely recognized for their ability to disinfect water, making it safe for drinking. Each has its advantages and considerations, making them essential tools in water purification.

Understanding Chemical Disinfection

Chemical disinfection works by killing pathogens in the water, including bacteria, viruses, and protozoa. The effectiveness of iodine and chlorine depends on the concentration of the chemical, water temperature, pH level, and contact time.

Using Iodine for Water Purification

Iodine has been a trusted method for water disinfection for decades. It is available in various forms, including tablets, solutions, and crystals.

- How to Use Iodine:
  - Tablets: Follow the manufacturer's instructions, typically one tablet per quart or liter of water.
  - Solution: Use about 5 drops of 2% iodine solution per quart or liter of water.
  - Crystals: Dissolve a few crystals in water to create a stock solution, then add the stock solution to the water being treated.

- Advantages:
  - Effectiveness: Kills most bacteria and viruses.
  - Portability: Small and easy to carry, making it ideal for emergency kits and backpacking.

- Considerations:
  - Taste: Iodine can leave an unpleasant taste, which can be mitigated by letting the treated water sit open to the air for a few hours or by adding vitamin C after the treatment period.
  - Effectiveness Against Cryptosporidium: Less effective against this particular protozoa.
  - Health Concerns: Not recommended for pregnant women, those with thyroid problems, or continuous use for more than a few weeks at a time.

Using Chlorine for Water Purification

Chlorine, particularly in the form of bleach or purification tablets, is another widely used chemical for water disinfection.

- How to Use Chlorine:
  - Bleach: Add 2 drops of regular, unscented household bleach (5-6% sodium hypochlorite) per quart or liter of water. Stir and let stand for 30 minutes.
  - Tablets: Follow the instructions on the package, as concentrations can vary.

- Advantages:
  - Availability: Household bleach is readily available and inexpensive.
  - Effectiveness: Efficiently kills bacteria, viruses, and many types of protozoa.

- Considerations:
  - Taste and Odor: Chlorine can leave a noticeable taste and odor, which can be reduced by letting treated water sit uncovered.
  - Chemical Sensitivity: Some individuals may be sensitive to chlorine.
  - Storage and Shelf Life: Chlorine tablets have a shelf life, and bleach degrades over time, especially if exposed to heat or sunlight.

Safety Precautions and Tips

When using chemical treatments, certain precautions and tips can ensure safety and effectiveness:

- Follow Directions: Adhere strictly to the recommended dosages and wait times.
- Water Temperature: Warmer water reacts better with chemicals. In cold conditions, increase the contact time.
- Turbidity: Pre-filter water to remove particulates, as turbidity can shield pathogens from chemical contact.
- Storage: Store chemicals in a cool, dark place to maintain their effectiveness.
- Alternate Methods: Consider combining chemical treatment with other purification methods, such as filtration, for enhanced safety.

Special Considerations

Understanding the limitations and special considerations of chemical treatments is crucial for effective use:

- Resistant Pathogens: Some pathogens, like Cryptosporidium, are resistant to chemical disinfectants, especially iodine.
- Long-term Use: Prolonged use of iodine and chlorine for water treatment can have health implications. It's advisable to use these methods in the short term or rotate with other purification methods.
- Allergies and Sensitivities: Be aware of any personal or group sensitivities to iodine or chlorine before choosing these as your primary treatment methods.

Chemical water treatment using iodine and chlorine provides a portable, effective solution for purifying water in a variety of situations. Whether you are hiking in the backcountry, preparing for emergencies, or traveling in areas where water quality is questionable, understanding how to use these chemicals safely and effectively can be a lifesaver. Remember, the ultimate goal is to ensure access to safe drinking water, and chemical treatment is a valuable tool in achieving that objective. As we progress through "Water Purification Essentials," the next chapter will explore DIY and natural filtration systems, offering more options to ensure you have the cleanest water possible in any situation.

Chapter 4: DIY and Natural Filtration Systems

In our journey through "Water Purification Essentials," we've explored finding safe water sources and various methods to purify it. However, in many situations, especially when dealing with turbid water, filtration becomes a necessity before any chemical treatment or boiling. Filtration can remove particulates, some microorganisms, and even improve the taste and smell of water. This chapter focuses on DIY and natural filtration systems that are effective, sustainable, and can be made with materials often found in nature or around the home.

The Basics of Water Filtration

Water filtration involves passing water through materials that trap and remove particles. Effective filtration systems can range from simple cloth filters to more complex arrangements using sand, charcoal, and gravel. The key to successful DIY filtration lies in understanding the materials and methods that can best remove contaminants from water.

Simple Cloth Filtration

- Materials Needed: Any clean cloth, cotton, or bandana.
- Method: Place the cloth over a container and pour water through it. This basic method removes large particulates.
- Applications: Ideal for pre-filtering before using more detailed purification methods.

Charcoal-Based Filtration System

Activated charcoal is excellent for removing certain chemicals, improving taste and odor, and even clearing some bacteria.

- Materials Needed: Charcoal from a fire, a container (like a plastic bottle), sand, gravel, and cloth.
- Building Steps:
  - Create a layer of cloth at the bottom of the container.
  - Add a layer of activated charcoal. If using charcoal from a fire, ensure it's completely cooled and crushed into small pieces.
  - Add layers of sand and gravel above the charcoal to trap larger particles.
  - Pour water through the setup, collecting it in another container below.

Sand and Gravel Filtration System

For removing sediment and some pathogens, a sand and gravel system is effective.

- Materials Needed: Sand, gravel, a container, and cloth or coffee filters.
- Building Steps:
  - At the bottom of the container, place a layer of cloth or a coffee filter.
  - Add a layer of fine sand, followed by a layer of coarse sand.
  - Top with a layer of gravel.
  - Slowly pour water through the layers, collecting the filtered water in another container.

Bamboo Filtration

Bamboo is a natural material that can be used to create a simple, effective water filter, leveraging its hollow structure.

- Materials Needed: A section of bamboo, sand, and small stones.
- Method: Fill the bamboo section with alternating layers of sand and small stones. Pour water through the bamboo, allowing it to drip out the other end.

Natural Plant-Based Filtration

Certain plants and their parts can be used for water filtration due to their porous and absorbent nature.

- Materials: Peat, moss, or cattails.
- Method: Construct a filter using these materials as the medium. Water passed through these materials can have particulates absorbed by the plant fibers.

Tips for Effective DIY Filtration

- Multiple Filtration Stages: Combining several of these methods can increase effectiveness.
- Regular Maintenance: Replace materials in your filter regularly to maintain its effectiveness.

- Pre-Filtering: Use a cloth filter to remove large debris before water passes through finer filtration systems.

Safety and Considerations

- Post-Filtration Treatment: Filtration does not remove all microorganisms. Always boil or chemically treat water after filtration to ensure safety.
- Material Safety: Ensure that all materials used, especially charcoal, are free from chemicals.

Advantages of DIY and Natural Filtration Systems

- Accessibility: Can be built with readily available materials.
- Cost-Effective: Eliminates the need for expensive, store-bought systems.
- Sustainability: Utilizes natural and reusable resources, reducing environmental impact.

DIY and natural filtration systems provide an accessible and efficient way to improve water quality, especially in survival situations or for those seeking to live sustainably. These systems, while invaluable in removing particulates and improving the aesthetics of water, are part of a comprehensive approach to water purification. Always follow filtration with a method guaranteed to kill pathogens, such as boiling or chemical treatment, to ensure the water is safe for consumption. As we conclude "Water Purification Essentials," remember that knowledge and preparedness are key to ensuring access to clean, safe water in any situation. By understanding and implementing these fundamental water purification and filtration techniques, you can secure a vital resource essential for health and survival.

Book 3:
Foraging for Survival

Chapter 1: The Forager's Guide to Plant Identification

Foraging for wild plants is an ancient practice that connects us to the land, offering a sustainable way to gather food, medicine, and materials. However, the ability to identify plants accurately is fundamental to foraging safely and responsibly. This chapter serves as a comprehensive guide to the skills, tools, and knowledge needed for effective plant identification, a cornerstone of foraging for survival.

Understanding the Basics of Plant Identification

Plant identification is both an art and a science. It requires observation, knowledge, and often, a bit of detective work.

- Observation Skills: Learn to observe the details of plants—leaves, stems, flowers, fruits, and roots. Note their colors, shapes, sizes, and textures.
- Botanical Terminology: Familiarize yourself with basic botanical terms to describe plant parts accurately. This language is universal among foragers, gardeners, and botanists.
- Field Guides: Invest in a good regional field guide that includes photographs and descriptions of native plants. Digital apps can also be valuable tools for identification.

Tools for Plant Identification

Several tools can enhance your ability to identify plants accurately:

- Field Guide or App: Choose a guide or app specific to your region for the most relevant information.
- Magnifying Glass: Helps examine small plant features, such as leaf hairs or flower parts.
- Notebook and Pen: Record observations and sketches of plants for future reference.
- Camera: Take photos of plants for further research or consultation with experts.

The Importance of Scientific Names

While common names are easier to remember, they can be misleading due to regional variations. Scientific names, on the other hand, are universally recognized.

- Binomial Nomenclature: Consists of two parts - the genus and the species. This system avoids the confusion that can arise from common names.
- Learning Scientific Names: While it may seem daunting, learning the scientific names of plants can greatly enhance your foraging knowledge and accuracy.

Identifying Plant Parts

Understanding the parts of a plant is crucial for identification and knowing which parts are edible or have medicinal properties.

- Leaves: Note their arrangement (alternate, opposite, whorled), shape, edge (smooth, toothed, lobed), and attachment.
- Flowers: Observe the color, size, shape, and number of petals. The arrangement of flowers on the plant can also be a key identifier.
- Stems and Roots: Examine the stem's texture (smooth, hairy) and the root system (taproot, fibrous). Some plants are identified by their unique stems or roots.
- Fruits and Seeds: The type of fruit (berry, nut, pod) and the seed's appearance can help identify the plant and indicate ripeness.

Seasonal Changes and Plant Identification

Plants change throughout the seasons, affecting their identifiability.

- Seasonal Variability: Recognize that plants may look different depending on the time of year. Some may only be identifiable by their flowers in spring or their fruit in autumn.
- Phenology: The study of seasonal changes in plants can aid in identification. Knowing when certain plants flower or fruit can narrow down possibilities.

Ethical Foraging and Identification

Ethical foraging ensures sustainability and respect for the environment.

- Positive Identification: Never consume a plant unless you are 100% sure of its identity. When in doubt, leave it out.
- Sustainable Harvesting: Take only what you need and leave enough behind for the plant to regenerate and for wildlife to use.
- Respect Private Property: Always seek permission before foraging on private land.

Common Mistakes in Plant Identification

Avoid common pitfalls that can lead to misidentification and potentially dangerous mistakes.

- Relying Solely on Photos: Photos can be misleading due to variations in lighting and plant conditions. Always cross-reference with a reliable guide.
- Overlooking Habitat: Many plants have specific habitats. Noting the environment where a plant is found can aid in identification.
- Ignoring Similar Species: Be aware of lookalikes, especially those that are toxic. Learn the distinguishing features that separate edible plants from their dangerous counterparts.

Practical Exercises for Improving Identification Skills

- Field Trips: Regularly visit different habitats to practice identifying plants in the wild. Note how plants vary between locations.
- Join Foraging Groups: Learning from experienced foragers can enhance your skills and knowledge.
- Create a Plant Journal: Document your findings, including descriptions, photos, and notes on habitats and uses. This personal reference can be invaluable over time.

Mastering plant identification is essential for any forager. It requires patience, practice, and a keen eye for detail. By developing your identification skills, you equip yourself with the ability to safely and sustainably harvest the bounty that nature provides. Remember, every plant has its story, and learning to read those stories through their leaves, flowers, and habitats opens up a world of foraging possibilities. As you grow in your foraging journey, let curiosity and respect for nature be your guides, ensuring that foraging remains a sustainable practice for generations to come.

Chapter 2: Regional Edible Plants and Their Discovery

The world around us is abundant with edible plants, each region offering its unique bounty. This chapter delves into the exploration and identification of regional edible plants, providing foragers with the knowledge to unlock the nutritional potential of their local ecosystems. Understanding regional flora not only enhances our foraging skills but also connects us deeper with our natural environment.

Understanding Regional Flora

The first step in discovering regional edible plants is to understand the ecosystems and climates of your area:

- Climate Zones: Recognize the climate zone you are in (temperate, tropical, arctic, etc.) as it significantly influences the types of plants available.
- Ecosystems and Habitats: From forests and meadows to wetlands and deserts, different environments host different species.
- Seasonal Variations: Be aware of the seasonal changes that affect plant availability and edibility.

Researching Edible Plants

A thorough investigation is crucial to safely foraging for edible plants:

- Field Guides and Resources: Use regional field guides, websites, and apps dedicated to local flora. These resources often provide detailed descriptions, photographs, and uses of plants.
- Local Foraging Experts: Engage with local foraging groups, workshops, or classes led by experienced foragers. Their knowledge of local edible plants can be invaluable.
- Ethnobotanical Studies: Research the historical uses of plants by indigenous and local communities, which can offer insights into edible and medicinal plants.

Common Edible Plants by Region

While each region boasts its unique flora, some widely found edible plants include:

- Temperate Regions:

- Dandelions (Taraxacum officinale): Leaves, roots, and flowers are edible.
  - Wild Garlic (Allium ursinum): Known for its distinctive smell and taste.
  - Nettles (Urtica dioica): Young leaves are edible when cooked.

- Tropical Regions:
  - Banana (Musa spp.): Not just the fruit but also the flowers and young shoots are edible.
  - Papaya (Carica papaya): Both the fruit and seeds are edible.
  - Bamboo (Bambusoideae): Young shoots are edible.

- Arid/Desert Regions:
  - Prickly Pear Cactus (Opuntia spp.): Fruits and young pads are edible.
  - Agave (Agave spp.): The sap, base, and young flowers are edible.
  - Mesquite (Prosopis spp.): Pods can be ground into flour.

- Arctic and Sub-Arctic Regions:
  - Cloudberries (Rubus chamaemorus): Berries are edible.
  - Fireweed (Chamerion angustifolium): Young shoots and leaves are edible.
  - Arctic Raspberry (Rubus arcticus): Berries are edible.

Foraging Ethics and Sustainability

Responsible foraging ensures the preservation of wild plant populations and their habitats:

- Take Only What You Need: Harvest in a way that allows plants to regenerate.
- Respect Wildlife: Remember that humans are not the only ones who depend on these plants.
- Leave No Trace: Minimize your impact on natural habitats during foraging excursions.

Safety Precautions

Safety is paramount when foraging for edible plants:

- Positive Identification: Confirm the identity of a plant using multiple sources before consuming it.
- Avoid Contaminated Areas: Steer clear of plants from polluted areas or those exposed to pesticides and herbicides.
- Allergy Test: If trying a plant for the first time, conduct a small patch test to check for allergic reactions.

Documenting Your Discoveries

Keeping a record of your foraging discoveries can enhance your knowledge and skills:

- Foraging Journal: Maintain a journal of the plants you discover, including notes on their location, appearance, and uses.
- Photographs: Take photographs of the plants in various stages of growth for future identification.
- Sample Preservation: Pressing small samples of plants can help in creating a personal reference guide.

Engaging with the Community

Sharing knowledge and experiences enriches the foraging community:

- Foraging Walks and Workshops: Participate in or organize local foraging walks to learn from and teach others.
- Online Forums and Social Media: Engage with online communities dedicated to foraging and wild edibles.
- Contribute to Citizen Science: Participate in projects that track and document local flora.

Discovering regional edible plants is a journey of exploration, learning, and connection with nature. It requires diligence, respect for the environment, and a commitment to sustainability. By developing a deep understanding of the edible plants specific to your region, you equip yourself with the knowledge to forage safely and sustainably. This chapter has laid the groundwork for recognizing and utilizing the edible bounty that surrounds us, fostering a deeper appreciation for the natural

Chapter 3: Avoiding Toxic Plants and Dangerous Lookalikes

Foraging for survival connects us with the wilderness, offering nourishment and healing through nature's bounty. However, the natural world is also home to plants that can harm us. This chapter delves into the critical skills needed to navigate the risks of toxic plants and their dangerous lookalikes, ensuring safe foraging practices.

Understanding Plant Toxicity

Toxic plants contain substances harmful to humans, which can cause reactions ranging from mild irritation to severe poisoning or even death. Understanding the types of toxins and their effects is the first step in safe foraging:

- Irritants: Cause skin rashes or digestive upset.
- Neurotoxins: Affect the nervous system, potentially leading to paralysis or neurological damage.
- Carcinogens: Possess cancer-causing properties.
- Organ Toxins: Affect specific organs such as the liver, kidneys, or heart.

Identifying Common Toxic Plants

Familiarity with common toxic plants in your region is essential. While these vary globally, some widely found toxic species include:

- Poison Ivy, Oak, and Sumac: Cause severe skin irritation.
- Foxglove (Digitalis purpurea): Contains heart-affecting glycosides.
- Deadly Nightshade (Atropa belladonna): Contains atropine, causing delirium and potentially death.

Dangerous Lookalikes

Many edible plants have toxic lookalikes, making accurate identification crucial. For example:

- Wild Carrot (Daucus carota) and Poison Hemlock (Conium maculatum): Poison hemlock has a smooth, hollow stem with purple spots, unlike the hairy stem of wild carrot.
- Morels and False Morels (Gyromitra spp.): True morels are hollow when cut open, while false morels contain cotton-like fibers inside.

Key Identification Techniques

Avoiding toxic plants and their lookalikes requires meticulous attention to detail:

- Learn Identifying Features: Study the distinguishing characteristics of plants, such as leaf patterns, stem textures, and flower colors.
- Use Multiple Resources: Consult various field guides, apps, and experts to confirm a plant's identity.
- Attend Workshops: Hands-on learning with experienced foragers can significantly improve your identification skills.

Foraging Best Practices

Implementing best practices can minimize the risks associated with toxic plants:

- Never Assume: Never consume a plant based on assumption. If in doubt, leave it out.
- Forage with Experts: When starting out, forage with knowledgeable individuals who can guide you.
- Be Cautious with New Plants: Introduce new plants into your diet gradually to monitor potential reactions.

The Role of Smell and Taste Tests

While smell and taste tests are traditional methods in plant identification, they must be approached with caution:

- Smell: Some toxic plants have a distinctive unpleasant odor, but not all. Use smell as a supplementary cue, not a definitive test.
- Taste: A tiny taste, immediately spit out, can sometimes help differentiate species. However, this should only be done under expert guidance and with plants that are already highly suspected to be safe.

Learning from History and Folklore

Historical and cultural knowledge can provide insights into plant safety:

- Ethnobotanical Records: Indigenous knowledge and historical records often contain valuable information on plant uses and dangers.

- Folklore: While not always scientifically accurate, folklore can sometimes offer warnings about certain plants.

Documenting and Reporting

Keeping a personal log of plant encounters enhances learning and can contribute to community knowledge:

- Photograph and Note: Document plants you encounter, noting location, appearance, and any identification efforts.
- Share Findings: Sharing your experiences with foraging communities online or in person can help spread knowledge about toxic plants and their lookalikes.

Handling and First Aid for Plant Poisoning

Knowing basic first aid for plant poisoning is crucial for every forager:

- Immediate Actions: If ingestion of a toxic plant is suspected, seek medical help immediately. Do not induce vomiting unless advised by a poison control center or medical professional.
- Skin Contact: If a skin irritant plant is touched, wash the area thoroughly with soap and water.

The ability to distinguish between edible and toxic plants is a vital skill in foraging for survival. This chapter has equipped you with the knowledge to navigate the green world safely, emphasizing the importance of careful identification, respect for nature's complexity, and continuous learning. By adhering to the principles of cautious foraging, you can enjoy the abundant gifts of the earth without falling foul of its defenses. Remember, the natural world is a generous provider but demands respect and wisdom in return. As we move forward to explore the best practices in harvesting and preparation, keep in mind the lessons learned from avoiding toxic plants and their lookalikes, ensuring a safe and rewarding foraging experience.

Book 4:
Hunting and Trapping Skills

Chapter 1: Beginner's Guide to Basic Hunting Techniques

Entering the realm of hunting is to step into an ancient tradition that connects us deeply with nature and our primal selves. This chapter is designed as a primer for those new to this venerable practice, covering essential hunting techniques that have been honed over generations. Whether your aim is to supplement your diet, become more self-sufficient, or engage with the outdoors in a profoundly interactive manner, mastering these basic techniques is your starting point.

Understanding Your Objectives

Before venturing into the field, it's crucial to clarify your hunting objectives. Are you targeting small game like rabbits and squirrels, or are you after larger prey such as deer? Your goals will dictate the techniques, tools, and preparation required.

The Right Tools for the Task

Selecting the appropriate weapon is your first practical step. Bows and firearms are the most common choices, each with its advantages and learning curve.

- Bows: Offer a silent approach and the satisfaction of mastering a challenging skill. Suitable for those who value stealth and tradition.
- Firearms: Provide a longer range and greater power. They're diverse, from shotguns suitable for small, fast-moving game to rifles designed for precision at a distance.

Safety First

- Education: Complete a hunter education course to learn about safety, wildlife laws, conservation, and ethics.

- Preparation: Familiarize yourself with your weapon. Practice regularly to ensure you can use it safely and effectively.
- Safety Gear: Always wear appropriate safety gear, including blaze orange during rifle seasons, to make yourself visible to other hunters.

Basic Hunting Techniques

1. Spot and Stalk

This method involves locating your prey from a distance and then stealthily moving into a position where you can make a clean, ethical shot.

- Patience and Observation: Spend time observing animal movements and behaviors. Early morning or late afternoon are prime times for many species.
- Stealth: Move slowly, use natural cover, and be mindful of wind direction to avoid alerting your prey with your scent.

2. Still Hunting

Still hunting is the practice of moving slowly through an area where game is likely to be found, stopping frequently to watch and listen.

- Silent Movement: Take slow, deliberate steps. Use soft ground cover to muffle sounds.
- Heightened Awareness: Use all your senses. Listening for sounds and looking for signs of movement can lead you to your target.

3. Ambush Hunting (Blind and Stand Hunting)

Ambush hunting involves waiting in a concealed position for game to come to you. Blinds or stands can be used for camouflage.

- Location: Choose a spot near game trails, water sources, or feeding areas where animals are likely to pass.
- Camouflage: Use natural vegetation or a purpose-built blind to conceal your presence.

4. Calling and Decoys

Using calls and decoys can attract game into range. This technique requires knowledge of animal behavior and sounds.

- Calls: Learn to mimic the sounds of animals. There are calls for almost every species, from ducks to deer.
- Decoys: Place decoys in visible locations to attract curious game. Ensure they are realistic and appropriate for the species and season.

The Importance of Scouting

Scouting the area where you plan to hunt is crucial. Look for signs of animal activity, including tracks, droppings, bedding areas, and feeding sites. Familiarity with the terrain and animal patterns will significantly increase your chances of success.

Ethical Considerations

Ethical hunting is about respect—for the animal, the environment, and other people. Always strive for a clean, humane kill to prevent unnecessary suffering. Know your target and what lies beyond it to avoid accidents.

The Role of Patience and Perseverance

Patience is perhaps the most critical skill in a hunter's arsenal. You may spend hours, even days, without seeing game. Perseverance, the willingness to learn from mistakes and persist despite setbacks, is equally vital.

Learning from Each Experience

Every outing is an opportunity to learn. Reflect on what worked, what didn't, and how you can improve. Engage with the hunting community to share stories and advice.

Hunting is a journey that offers endless lessons in nature, wildlife, and self-reliance. As a beginner, your focus should be on learning the basics, respecting the ethics of the hunt, and continuously improving your skills. Remember, hunting is not just about the harvest but about the experience, the connection to the natural world, and the ancient lineage of hunters that you're now a part of. With dedication, patience, and respect, you'll find that hunting can enrich your life in ways you never imagined.

Chapter 2: Constructing Effective Traps for Game

Trapping is an age-old practice that serves as a crucial survival skill and an effective method for securing food. This chapter explores the art of constructing effective traps for game, offering insights into the mechanics, design, and ethical considerations of trapping. Whether for survival situations, wildlife management, or sustainable living, understanding how to build and deploy traps responsibly is essential.

Understanding Trapping

Trapping involves setting devices or contraptions designed to catch or hold game. It's a skill that requires patience, ingenuity, and a deep understanding of animal behavior and habitat.

Types of Traps

Several types of traps can be used, depending on the target species and environment:

- Snares: A looped wire or cord designed to tighten around the animal.
- Deadfalls: A heavy weight that falls to trap the animal beneath it.
- Cage Traps: Enclosures that capture the animal alive.
- Pitfalls: Holes dug into the ground, concealed to capture animals that fall into them.

Essential Considerations for Trapping

- Local Laws and Regulations: Always check and adhere to local laws regarding trapping. Some methods are restricted or prohibited in certain areas.
- Target Species: Tailor your trap to the specific game you're targeting. Consider the animal's size, behavior, and habitat.
- Safety: Be mindful of the safety of other people, pets, and non-target wildlife. Place traps in areas where they are unlikely to cause unintended harm.

Constructing a Snare Trap

Snares are simple yet effective for small to medium-sized game. Here's how to construct a basic snare:

1. Materials: Use strong, flexible wire or high-tensile strength cordage.
2. Location: Set snares on known animal paths or near burrows and watering holes.
3. Loop Size: Adjust the loop size according to the target species. A rabbit, for example, requires a loop about the size of your hand.
4. Anchoring: Securely anchor the snare to a solid object like a tree or stake in the ground.
5. Height: Position the loop off the ground at the height of the target animal's head.

Building a Deadfall Trap

Deadfalls use weight to trap an animal, suitable for small game:

1. Materials: You'll need a heavy rock or log and triggering sticks.
2. Structure: Arrange the sticks in a four-figure pattern or use a simple lever and fulcrum design to support the weight.
3. Bait: Place bait on the trigger mechanism to entice the animal.
4. Sensitivity: Adjust the trigger to be sensitive enough that the weight falls when the bait is disturbed but not so sensitive that it's triggered by wind or small non-target animals.

Creating a Cage Trap

Cage traps are humane and allow for the release of non-target animals:

1. Materials: Wire mesh or wooden slats can be used to construct the cage.
2. Door Mechanism: The door should close securely when the animal triggers the mechanism, usually by stepping on a pressure plate or moving the bait.
3. Size and Bait: The cage size should match the target species, and appropriate bait should be used to lure the animal inside.

Pitfall Traps

Pitfalls are passive traps that rely on animals falling into a concealed hole:

1. Digging: Choose a location on an animal path and dig a hole deep enough that the animal cannot climb out.
2. Concealment: Cover the hole with branches, leaves, or other natural materials.
3. Warning: Be extremely cautious with pitfall traps, as they can be dangerous to other animals and humans if not properly marked or monitored.

Ethical Trapping Practices

Ethical considerations are paramount in trapping. The goal is to ensure the humane treatment of animals and minimal impact on the environment:

- Check Traps Regularly: This reduces the suffering of captured animals and allows for the timely release of non-target species.
- Use Humane Killing Methods: If the intent is to kill, it should be done quickly and humanely to minimize suffering.
- Conservation Mindset: Trap with conservation in mind, ensuring that trapping activities do not negatively impact local wildlife populations.

Constructing effective traps for game is a skill that marries technical knowledge with ethical responsibility. It requires understanding the delicate balance of nature and the impact of human intervention. By mastering these techniques and adhering to ethical practices, trappers can ensure that they respect the wildlife and ecosystems they engage with. Trapping, when done correctly, can be a sustainable part of living closely with the natural world, providing food, fur, and an invaluable connection to our ancestral roots. As we move forward, remember that trapping is not just about the skillful capture of wildlife but about integrating respect, conservation, and sustainability into every action we take in the wild.

Chapter 3: Mastering the Art of Tracking and Stalking

The art of tracking and stalking is a testament to the hunter's skill, patience, and profound connection with the natural world. It is a practice that goes beyond the mere act of pursuit to encompass understanding, respect, and the ethical engagement with wildlife. This chapter delves into the nuances of tracking and stalking, offering insights into how to hone these essential skills for successful and ethical hunting.

The Essence of Tracking

Tracking is the ability to read signs left by animals, interpret their movements, and predict their behavior. It's an ancient skill that requires keen observation, knowledge of animal habits, and an understanding of the environment.

- Signs to Look For: Tracks, scat, trails, bedding areas, feeding signs, and rubbings on trees. Each of these signs can tell a story about the animal's size, health, direction, and how recently it passed.
- Ground Sign Awareness: Learn to recognize disturbances on the ground, differences in vegetation, and subtle changes in the environment that may indicate animal activity.
- Understanding Animal Behavior: Knowledge of your target species' habits, preferred foods, and patterns of movement is crucial for successful tracking.

Techniques for Effective Tracking

- Start Slow: Begin in an area known for animal activity. Practice identifying and following tracks, even if you do not plan to hunt.
- Use the Right Tools: A good pair of binoculars, a field guide to animal tracks, and a notebook for recording observations can be invaluable.
- Practice in Different Terrains: Tracking skills are sharpened by experience in various environments—from dense forests to open plains.

The Strategy of Stalking

Stalking involves quietly and carefully moving closer to the game without detection. It's a skill that combines stealth, patience, and precision.

- Wind Direction: Always be aware of the wind direction to avoid your scent reaching the animal. Move with the wind in your face or from the side.
- Silent Movement: Learn to walk quietly, placing your feet gently and using vegetation to muffle your steps. Wear soft, quiet clothing to minimize noise.
- Visual Concealment: Use natural cover to camouflage your approach. Understand the concept of "breaking up your silhouette" to blend into the environment.

The Role of Patience

Patience is perhaps the most critical attribute in tracking and stalking. Waiting for the right moment to move or take a shot often makes the difference between success and failure.

- Observation: Spend time observing animals from a distance to learn their patterns and choose the best approach.
- Timing: Know when to move and when to stay still. Often, the key to successful stalking is knowing when not to advance.

Ethical Considerations in Tracking and Stalking

Ethical tracking and stalking are grounded in respect for the animal and the environment.

- Respect for the Quarry: Approach your target with the respect it deserves. Remember, you're engaging with a living creature.
- Intention: Only track and stalk with the intention of making a clean, ethical shot. Avoid actions that would unnecessarily stress or endanger the animal.
- Leave No Trace: Practice environmental stewardship by minimizing your impact on the natural surroundings.

Mastering the Art

Mastering tracking and stalking is a lifelong journey that evolves with experience.

- Continuous Learning: Engage in regular practice, seek knowledge from experienced hunters, and remain open to learning from each outing.

- Mindfulness and Connection: Develop a deeper connection with nature. Mindfulness and awareness are as important as physical skills in tracking and stalking.
- Resilience: Learn from unsuccessful attempts. Resilience and adaptability are key to refining your skills.

The art of tracking and stalking is a profound practice that enhances our engagement with the natural world. It teaches us not only about the wildlife we pursue but also about ourselves—our patience, our respect for nature, and our place within the ecosystem. As you continue to develop these skills, remember that the true measure of success in hunting is not always the game secured but the knowledge gained, the ethical approach maintained, and the connection to the wild deepened. Through mastering tracking and stalking, hunters can foster a sustainable, respectful, and deeply rewarding relationship with the natural world, ensuring that the tradition of hunting continues to be an ethical and honorable pursuit.

Chapter 4: Respecting Wildlife with Ethical Hunting Practices

In the realm of hunting and trapping, ethics stand as the cornerstone of respect for wildlife, ensuring the sustainability of ecosystems and the continuation of hunting traditions for future generations. This chapter delves into the principles of ethical hunting practices, emphasizing the responsibility of hunters to honor the life they pursue and the natural world they engage with.

The Foundation of Ethical Hunting

Ethical hunting is rooted in a deep respect for nature and wildlife, involving more than following laws and regulations. It encompasses a personal commitment to sustainable practices, humane methods, and the conservation of wildlife habitats.

- Fair Chase: Pursue game in a manner that allows the animal a fair chance to escape, avoiding practices that give undue advantage to the hunter.
- Taking Ethical Shots: Aim only for shots that are likely to result in a quick, humane kill, minimizing animal suffering.

Understanding Wildlife Conservation

Conservation is integral to ethical hunting. Hunters play a vital role in maintaining healthy wildlife populations and ecosystems through regulated hunting and contributions to conservation efforts.

- Supporting Conservation Efforts: Engage with and support organizations dedicated to wildlife conservation and habitat preservation.
- Sustainable Harvest: Adhere to quotas and season dates that are scientifically determined to ensure the sustainability of wildlife populations.

The Role of Education

A well-informed hunter is an ethical hunter. Education on species behavior, habitat, and management strategies is crucial for making informed decisions in the field.

- Continuous Learning: Stay informed about the latest research and management strategies related to the species you hunt.
- Hunter Education: Participate in hunter education programs that cover safety, ethics, wildlife laws, and conservation.

Ethical Treatment of Game

Respect for the animal extends beyond the hunt to how one handles and utilizes the game.

- Respectful Handling: Treat game with respect from the moment of harvest through processing and utilization.
- Waste Minimization: Make every effort to use as much of the animal as possible, minimizing waste and honoring the life taken.

The Importance of Land Stewardship

Ethical hunters recognize their role as stewards of the land, taking action to preserve natural habitats and biodiversity.

- Habitat Protection: Participate in or support habitat restoration projects and practices that protect the ecosystems relied upon by wildlife.
- Leave No Trace: Practice Leave No Trace principles while hunting, ensuring minimal impact on natural habitats.

Engaging with the Community

Ethical hunting practices are reinforced and spread through engagement with the hunting community and the public.

- Mentorship: Experienced hunters should mentor new hunters, instilling ethical practices and respect for wildlife.
- Public Engagement: Actively engage in dialogues about hunting and conservation, providing accurate information and dispelling misconceptions.

Ethical Decision-Making in the Field

Ethical hunting involves moment-to-moment decision-making in the field, guided by a moral compass focused on respect for life and nature.

- Self-Regulation: Even when laws may allow certain actions, ethical hunters consider the broader implications of their decisions on wildlife and ecosystems.
- Conflict Resolution: Address unethical behavior in others through constructive dialogue, emphasizing the importance of ethical practices for the future of hunting.

The Legacy of Ethical Hunting

The legacy of hunting is preserved through the ethical conduct of hunters today, ensuring that future generations can continue to engage with and learn from the natural world.

- Conservation Legacy: By prioritizing ethical practices, hunters contribute to a legacy of conservation and respect for wildlife that will endure for generations.
- Cultural Heritage: Ethical hunting practices honor the traditions of hunting and trapping, ensuring they are passed down with integrity and respect.

Ethical hunting practices are the hallmark of a responsible and respectful hunter. By adhering to principles of fair chase, conservation, education, and stewardship, hunters can ensure that their engagement with the natural world is sustainable, respectful, and aligned with the broader goals of wildlife preservation and ecosystem health. As we close this chapter on ethical hunting practices, let us carry forward the commitment to honor and respect wildlife, embracing our role not just as hunters but as guardians of the natural world. Through ethical practices, we not only enrich our own hunting experiences but also contribute to the health and vitality of ecosystems and the preservation of hunting traditions for future generations.

Book 5:
Fire Making Mastery

Chapter 1: The Science Behind Fire Basics

The mastery of fire is one of humanity's oldest and most significant achievements. It has been a cornerstone of our survival, providing warmth, protection, and a means to cook food. This chapter delves into the science behind fire basics, unraveling the principles that govern the creation and maintenance of fire. Understanding these principles is essential for anyone looking to develop or enhance their fire-making skills, whether for outdoor adventures, survival situations, or simply the joy of mastering one of nature's most powerful elements.

Understanding the Fire Triangle

At the heart of fire-making is the Fire Triangle, a simple model that explains the three essential elements a fire needs to ignite and sustain: Fuel, Heat, and Oxygen.

- Fuel: Acts as the combustible material. In fire-making, this could be wood, paper, or any organic matter capable of burning.
- Heat: Provides the energy required to start the combustion process. This can come from a spark, friction, or an external flame.
- Oxygen: Supports the chemical reactions that occur during combustion, typically sourced from the surrounding air.

The Combustion Process

Combustion is a chemical reaction between fuel and oxygen that releases heat and light. This process can be broken down into stages:

1. Pre-ignition: The fuel is heated to its ignition temperature, leading to the release of volatile gases in a process called pyrolysis.
2. Ignition: If the gases' temperature is high enough and in the presence of oxygen, combustion begins.
3. Combustion: The fire burns, sustained by the ongoing consumption of fuel and oxygen, and the removal of combustion products.

Understanding these stages is crucial for effective fire-making, as each stage requires specific conditions to proceed.

The Role of Heat in Fire-Making

Heat is the catalyst that starts the combustion process. There are several methods to generate heat for fire-making:

- Friction: The most ancient method, where heat is generated by rubbing two pieces of wood together.
- Percussion: Striking a hard object, like flint, against steel to create sparks.
- Concentration of Light: Using a lens to focus sunlight onto a small point, increasing its temperature.

Selecting the Right Fuel

Not all fuels are created equal. The choice of fuel can significantly affect the ease of starting a fire and its sustainability.

- Tinder: Fine, dry material that catches fire easily from a spark. Examples include dry leaves, grass, or cotton balls.
- Kindling: Small sticks or branches that catch fire from the tinder and help to build the fire up.
- Firewood: Larger pieces of wood that sustain the fire once it is established.

The moisture content, size, and type of fuel all play critical roles in the combustion process. Dry, dead materials are preferable, as moisture inhibits combustion by absorbing heat.

Oxygen and Ventilation

Oxygen is essential for sustaining a fire, and proper ventilation ensures a continuous supply. When building a fire:

- Structure: Construct your fire in a way that allows air to flow freely. Techniques such as the teepee, log cabin, or lean-to structures are effective.
- Management: As the fire burns, manage it by occasionally adjusting the structure to improve airflow and add fuel as needed.

Extinguishing Fires Safely

Understanding how to extinguish a fire is as important as knowing how to start one. Removing any element of the Fire Triangle will put out a fire:

- Smothering: Deprives the fire of oxygen. Sand, dirt, or a fire blanket can be used for smothering.
- Cooling: Reduces the temperature below the ignition point. Water is the most common cooling agent.
- Starvation: Involves removing the fuel source, though this is often less practical in immediate situations.

Environmental Considerations

When making a fire, especially in the wilderness, it's vital to consider the environmental impact:

- Leave No Trace: Use existing fire rings or make fires on surfaces that won't be damaged. Always fully extinguish fires and scatter cool ashes.
- Safety Precautions: Be aware of your environment and local fire regulations. Never leave a fire unattended and ensure it's completely out before leaving.

The science behind fire basics is both fascinating and practical. By understanding the Fire Triangle, the combustion process, and the roles of heat, fuel, and oxygen, anyone can become proficient in fire-making. This knowledge not only enhances one's ability to create fire when needed but also instills a deeper respect for this powerful element. As we move forward to explore modern fire starters, maintaining and safeguarding your fire, and creating emergency signal fires, remember that the mastery of fire begins with a solid understanding of its scientific principles. This foundation will serve as a guide through the art and science of fire-making, ensuring safety, efficiency, and environmental responsibility.

Chapter 2: Modern Fire Starters for Quick Lighting

In the quest for fire-making mastery, modern technology has provided us with an array of tools designed for quick and efficient fire lighting. These innovations not only simplify the process but also ensure reliability in a variety of conditions, from damp environments to high-altitude situations. This chapter explores the various modern fire starters available to outdoor enthusiasts, survivalists, and anyone interested in honing their fire-making skills with contemporary solutions.

Understanding Modern Fire Starters

Modern fire starters are designed to overcome traditional challenges associated with fire-making, such as wet conditions or lack of suitable natural tinder. They can be categorized into several types, each with its unique advantages and applications.

Ferrocerium Rods (Ferro Rods)

Ferrocerium rods, commonly known as ferro rods, are a popular choice among survivalists and outdoor adventurers. Made from a blend of metals that produce hot sparks when scraped, they work even when wet.

- Usage: Scrape the rod with a striker or the back of a knife blade to create sparks.
- Advantages: Durable, reliable in all weather conditions, and capable of producing thousands of sparks.
- Applications: Ideal for camping, survival kits, and emergency preparedness.

Magnesium Blocks

Magnesium blocks serve a dual purpose: a flammable magnesium shavings source and a sparking insert to ignite them.

- Usage: Shave small magnesium pieces onto your tinder pile and use the attached ferro rod to ignite the shavings.
- Advantages: Magnesium burns at an extremely high temperature, making it effective even with damp tinder.
- Applications: Useful in adverse weather conditions where traditional tinder fails to ignite.

Windproof and Waterproof Matches

These are not your ordinary matches; they are designed to withstand extreme conditions, ensuring a flame even in heavy wind or rain.

- Usage: Strike on the specially designed surface provided with the matches.
- Advantages: Easy to use, with a longer burn time than regular matches.
- Applications: Emergency kits, outdoor adventures, and situations where quick lighting is crucial.

Butane Lighters and Stormproof Lighters

Butane lighters, particularly those designed for outdoor use, offer a convenient fire source. Stormproof lighters take it a step further with windproof and waterproof capabilities.

- Usage: Operate with a push-button mechanism to produce a flame.
- Advantages: Instant flame, refillable options, and durable designs for outdoor use.
- Applications: Versatile for camping, grilling, and emergency situations.

Electric Arc Lighters

Electric arc lighters, or plasma lighters, use electricity to create a plasma arc hot enough to ignite a fire. They are rechargeable, typically via USB.

- Usage: Press a button to create an electric arc between electrodes.
- Advantages: Windproof, no fuel required, and environmentally friendly.
- Applications: Camping, survival kits, and as a backup fire source.

Solar Fire Starters

Harnessing the power of the sun, solar fire starters focus sunlight into a concentrated beam to ignite tinder.

- Usage: Position the solar starter to focus sunlight onto a piece of tinder until it catches fire.
- Advantages: Unlimited uses, no consumables, and works well in sunny conditions.

- Applications: Ideal for wilderness survival and educational purposes to demonstrate solar energy's power.

Chemical Fire Starters

Chemical fire starters include various compounds that ignite quickly and sustain a burn, even in challenging conditions.

- Usage: Apply a small amount to tinder and ignite.
- Advantages: Fast-acting, with a long shelf life and consistent performance.
- Applications: Emergency fire starting, quick campfire setup, and use in damp conditions.

Choosing the Right Fire Starter

When selecting a modern fire starter, consider the following factors:

- Environment: Choose a fire starter suited to your typical outdoor conditions (wet, windy, cold).
- Ease of Use: Ensure the fire starter is something you're comfortable using, especially under stress.
- Durability: Opt for robust, reliable options that won't fail when you need them most.
- Sustainability: Consider the environmental impact, opting for reusable or environmentally friendly options when possible.

The evolution of fire starters from traditional methods to modern innovations reflects humanity's ongoing relationship with fire. These modern tools not only offer convenience and efficiency but also enhance safety and reliability in fire-making. Whether for emergency preparedness, outdoor adventures, or simply mastering the art of fire, understanding and utilizing these modern fire starters can significantly elevate one's fire-making prowess. As we explore further into maintaining and safeguarding your fire and creating emergency signal fires, remember that the essence of fire mastery lies not just in the tools we use but in our respect for the flame and the knowledge we apply in harnessing its power.

Chapter 3: Maintaining and Safeguarding Your Fire

Once you've mastered the initial spark, the next crucial steps are maintaining your fire and ensuring it remains a safe, controlled element within your environment. This chapter focuses on the techniques and considerations necessary for keeping your fire burning efficiently and safeguarding against potential hazards. Proper fire management not only extends the utility and enjoyment of your fire but also underscores your responsibility as a steward of nature.

Maintaining Your Fire

Maintaining a fire requires attention to its structure, fuel supply, and the balance of the fire triangle elements: oxygen, heat, and fuel.

Fuel Management

- Gradual Addition: Add fuel in stages, starting with tinder and kindling, and gradually introducing larger pieces of wood. This method ensures a steady burn without smothering the flames.
- Fuel Types: Use a variety of fuel types to manage the fire's intensity. Hardwoods burn slower and are ideal for a long-lasting fire, while softwoods catch fire quickly, providing immediate heat.
- Consistent Supply: Keep a consistent supply of fuel nearby, ensuring it's dry and readily available to add as needed.

Airflow Regulation

- Structural Integrity: Construct your fire in a manner that promotes good airflow. Techniques such as the teepee, log cabin, or lean-to allow air to circulate freely, feeding the flames.
- Adjusting the Fire: Periodically reposition logs and embers to improve ventilation. This can help revitalize a dwindling fire and prevent it from going out prematurely.

Heat Conservation

- Reflectors: Position a backdrop of rocks or build a reflector wall with logs behind the fire to reflect heat back towards your camp or shelter.
- Windbreaks: Use natural land formations or build barriers to protect your fire from strong winds, which can quickly dissipate heat or spread flames uncontrollably.

Safeguarding Your Fire

While fire is a valuable resource, it poses significant risks if not properly managed. Safeguarding your fire involves preventive measures to avoid accidents and potential wildfires.

Fire Pit Selection and Preparation

- Location: Choose a location for your fire that's away from flammable materials like dry leaves, grass, or overhanging branches. Ensure it's downwind from your shelter or campsite to avoid smoke inhalation and fire spread.
- Clearing the Area: Clear a wide perimeter around your fire site of any debris, leaves, or other combustible materials. This barrier helps prevent the accidental spread of fire.
- Fire Pit Construction: Dig a shallow pit and surround it with rocks to contain the fire. This not only focuses the heat but also acts as a barrier against spreading.

Monitoring and Control

- Never Leave Unattended: A cardinal rule of fire safety is never to leave your fire unattended. Wind conditions can change, and embers can escape, leading to unintended fires.
- Water Supply: Always have water, sand, or dirt on hand to extinguish the fire quickly if it becomes necessary.
- Size Management: Keep the fire manageable. Larger fires can be more difficult to control and pose a greater risk of spreading.

Extinguishing Your Fire

Properly extinguishing your fire is as important as starting it. The goal is to ensure that the fire is completely out and poses no risk of reigniting.

- Douse with Water: Slowly pour water over the embers, stirring them to ensure all hot spots are cooled. Avoid pouring too much water too quickly, as this can create a hard crust over embers that retain heat underneath.
- Scatter Ashes: Once the embers are cool, scatter the ashes over a wide area, ensuring they are fully extinguished. Feel the ashes with the back of your hand to ensure there's no residual heat.
- Restore the Site: If you've created a fire pit, fill it back in and return the site to its natural state as much as possible.

Ethical Considerations

Ethical fire management extends beyond safety, encompassing a respect for the environment and future generations.

- Minimal Impact: Use established fire rings where available and adhere to Leave No Trace principles, minimizing your environmental impact.
- Consider Alternatives: In areas where fire impact is high or resources are scarce, consider using a portable stove or fire pan to reduce your footprint.

Fire making is an art that extends beyond the initial ignition to encompass the careful maintenance and ethical stewarding of this powerful element. By mastering the techniques of maintaining and safeguarding your fire, you ensure not only your safety and comfort but also the preservation of the natural beauty that surrounds us. Fire, when respected and managed with care, remains one of humanity's most invaluable tools, providing warmth, light, and a connection to our ancestral roots. As we move forward to explore creating emergency signal fires, remember that the principles of maintenance and safeguarding are foundational to all aspects of fire mastery, reflecting our responsibility towards both our survival and the environment

Chapter 4: Creating Emergency Signal Fires for Rescue

In a survival situation, being able to signal for rescue can mean the difference between life and death. An emergency signal fire is one of the most effective ways to attract the attention of rescuers, whether you're lost in the wilderness, stranded on a deserted island, or in any situation where you need to be found. This chapter focuses on how to create and use signal fires effectively, ensuring that when you need help, your fire will serve as a beacon that leads rescuers to you.

Understanding the Purpose of Signal Fires

Signal fires are not just any fires; they are specifically designed and located to catch the attention of search and rescue teams from the air or ground. The key is to create smoke and flames that are visible from a distance, making it clear that someone is in need of help.

Planning Your Signal Fire

Before you strike a spark, planning is crucial to maximize the effectiveness of your signal fire.

- Location, Location, Location: Choose a location that's visible from the air and from the ground in all directions if possible. Clearings, hilltops, or the edge of a water body can be ideal.
- Prepare in Advance: Even if you're not in immediate distress, preparing a signal fire site when you first realize you're lost can save precious time later.
- Signal Fire Triangle: Just like the fire triangle, remember the signal fire triangle: visibility, contrast, and sustainability. Your fire needs to be seen, stand out against its background, and last long enough to be noticed.

Constructing Your Signal Fire

A signal fire needs to be larger and produce more smoke than a standard campfire. Here's how to construct one:

The Basics

- Clear a Large Area: Remove all flammable material around the fire site to prevent unintended spreading.
- Create a Platform: Build a platform using green branches or logs to elevate the fire from wet ground and to hold more material.

For Smoke

- Materials for Smoke: Gather materials that produce thick smoke when burned. Green branches, leaves, and even certain types of rubber or oil can increase smoke production.
- Multi-Layer Approach: Start with a layer of dry material to ignite the fire quickly, followed by wet/green material to produce smoke.

For Visibility

- Nighttime Fires: For night use, focus on bright flames. Dry, resinous wood that produces bright flames can be more visible.
- Daytime Fires: For daytime, smoke is your signal. White smoke contrasts well against green or dark backgrounds, while dark smoke stands out against the sky or snow.

Lighting the Signal Fire

When it comes time to light your signal fire, timing and readiness are everything.

- Wait for the Right Moment: If you hear or see potential rescuers nearby, that's the time to light your fire to ensure it's noticed.
- Keep it Ready: Have your signal fire set up so that it can be lit quickly. Keep a separate fire or embers burning nearby to ignite your signal fire rapidly.

Maintaining and Maximizing Visibility

Once lit, maintaining your signal fire is critical to ensuring it continues to serve its purpose.

- Feed the Fire: Regularly add more green material to maintain smoke production during the day and dry material at night to keep flames visible.
- Multiple Fires: If possible, create three fires in a triangle or in a straight line with about 100 feet between them. This configuration is internationally recognized as a distress signal.

Ethical and Environmental Considerations

While your survival is paramount, it's important to balance this with ethical and environmental considerations.

- Minimize Damage: Use existing clearings or natural openings to reduce environmental impact.
- Extinguish Thoroughly: Once rescue is assured, or if you must leave the location, ensure your fire is completely extinguished to prevent wildfires.

Creating an effective emergency signal fire requires foresight, preparation, and knowledge. By understanding the principles outlined in this chapter and applying them with consideration for both visibility and safety, you can significantly increase your chances of being found in a survival situation. Signal fires are a powerful tool in your survival arsenal, embodying the hope for rescue and the human will to survive against the odds. Remember, the goal of mastering fire-making skills, especially in emergencies, is not just about survival but about doing so responsibly, with respect for the natural world that sustains us.

Book 6:
Survival Navigation

Chapter 1: Map-Free Navigation Tips and Tricks

In the wilderness, the ability to navigate without a map is a vital survival skill. Whether you're lost without a GPS, in a scenario where technology fails, or simply challenging yourself to rely on natural navigation methods, understanding the landscape and using the earth's natural cues can guide your path. This chapter delves into map-free navigation tips and tricks, empowering you to find your way using the wisdom of ancient travelers and the innate tools provided by nature.

Understanding Natural Navigation

Natural navigation involves interpreting the natural environment to determine directions and make informed decisions about your route. This ancient practice relies on observations of the sun, moon, stars, vegetation, landforms, and even animal behavior.

The Sun: Your Daily Guide

The sun is perhaps the most straightforward celestial body to use for navigation. Its predictable path can help orient you during daylight hours.

- Sunrise and Sunset: The sun rises in the east and sets in the west, providing a basic east-west orientation.
- Shadow Stick Method: Place a stick vertically in the ground and mark the shadow's tip. Wait 15-30 minutes and mark the new position of the shadow's tip. Drawing a line between the two marks gives you an east-west line, with the first mark being west.
- Watch Method: With an analog watch, point the hour hand at the sun. Bisect the angle between the hour hand and the 12 o'clock mark to find the south in the Northern Hemisphere (north in the Southern Hemisphere).

Using the Stars for Nighttime Navigation

The night sky is a map of its own, with constellations serving as guides for centuries.

- The North Star (Polaris): In the Northern Hemisphere, Polaris reliably indicates true north. Find the Big Dipper and follow the line drawn from its two outer stars upward to the next bright star.
- The Southern Cross: In the Southern Hemisphere, extend the long axis of the Southern Cross four and a half times its length to point south.

Interpreting Vegetation and Landforms

The natural world provides subtle clues to direction, influenced by sunlight, moisture, and prevailing winds.

- Moss and Tree Growth: Moss tends to grow on the damper, shadier side of trees, often the northern side in the Northern Hemisphere. However, use this method with caution as it's not universally reliable.
- Tree Branches: In general, tree branches tend to be denser and more horizontal on the southern side (in the Northern Hemisphere) due to greater sun exposure.
- Slope Orientation: In mountainous or hilly terrain, slopes facing the sun (south in the Northern Hemisphere, north in the Southern Hemisphere) often have less vegetation due to higher exposure to the sun.

Animal Behaviors as Indicators

Animals, insects, and birds can also provide navigation clues through their behavior and habitats.

- Bird Flight: Birds often fly towards water bodies in the evening and away in the morning.
- Ant Hills: In some regions, ants build their mounds on the warmer, sunnier side of trees or hills.
- Animal Trails: Trails tend to run north-south along ridges and east-west along valleys and water sources.

The Importance of Situational Awareness

Effective map-free navigation requires a heightened sense of awareness and the ability to read subtle environmental signs.

- Landmark Identification: Use distinctive landforms as reference points to maintain orientation and track progress.
- Mental Mapping: Continuously build a mental map of your surroundings, noting the direction and distance traveled.

- Backtracking: Leave markers or create easily recognizable signs to enable safe backtracking if needed.

Combining Techniques for Accuracy

No single navigation method is foolproof, especially in diverse terrains and climates. Combining several techniques increases reliability.

- Cross-Referencing: Use multiple natural indicators to confirm directions, such as combining sun position with wind direction and terrain features.
- Practicing: Regular practice in different environments enhances your skill and confidence in natural navigation.

Psychological Aspects of Map-Free Navigation

Staying calm and maintaining a positive mindset are crucial in navigation scenarios, especially in survival situations.

- Avoid Panic: Take deliberate, thoughtful actions to prevent disorientation and rash decisions.
- Trust Your Skills: Confidence in your navigation abilities can be as important as the skills themselves.

Ethical Considerations and Leave No Trace

Responsible navigation includes minimizing your impact on the environment.

- Leave No Trace: Practice minimal impact techniques to preserve the natural and cultural integrity of the places you explore.
- Respecting Wildlife: Be mindful of wildlife habitats and sensitive areas, navigating in a way that avoids disturbance.

Navigating without a map is a liberating skill that connects you with the natural world in a profound way. It hones your observational skills, deepens your understanding of the environment, and instills a sense of self-reliance.

Chapter 2: Crafting Improvised Compasses for Direction

In the wilderness, the ability to determine direction can be a lifeline. While the sun, stars, and natural landmarks offer invaluable guidance, sometimes the precision of a compass is irreplaceable. But what happens when you find yourself without this crucial tool? This chapter explores the ingenuity of crafting improvised compasses, enabling you to find your way even when traditional navigational tools are out of reach.

The Science Behind a Compass

At its core, a compass operates on the principle of magnetism. The Earth itself is a giant magnet, with magnetic fields stretching from the South Pole to the North Pole. A compass needle, which is magnetized, aligns itself with these fields, pointing towards magnetic north. Understanding this fundamental principle is the first step in crafting your own compass.

Materials and Methods

Creating an improvised compass requires minimal materials, many of which can be found in a natural setting or are common items in a survival kit.

Water and Leaf Compass

One of the simplest methods to create an improvised compass involves water, a leaf, and a piece of metal that can be magnetized (such as a needle).

- Magnetize the Needle: Stroke the needle in one direction with silk, wool, or your hair to magnetize it. If you have a silk or synthetic fabric, this can also work.
- Prepare the Water: Fill a container with still water. If no container is available, dig a small hole and line it with a plastic bag or leaves to hold water.
- Float the Needle: Place the needle on a leaf or a piece of paper, ensuring it doesn't sink, and gently set it on the water's surface. The needle will align itself with the Earth's magnetic fields, pointing towards magnetic north.

Shadow-Stick Compass

This method utilizes the sun and shadows to determine direction, bypassing the need for magnetism.

- Find a Stick: Place a stick vertically in the ground so that it casts a visible shadow.
- Mark the Shadow Tip: Use a stone or any object to mark the tip of the shadow. Wait about 15-30 minutes.
- Mark the New Position: Mark the new position of the shadow tip. The line between the first and second marks runs east-west, with the first mark west and the second mark east.
- Determine North: Standing with the first mark (west) to your left and the second mark (east) to your right, you are now facing true north.

Magnetizing with Electricity

If you have access to a battery, you can magnetize a needle more effectively.

- Use a Battery: By rubbing a needle using a battery (connecting the needle with the positive and negative ends for a few minutes), you can magnetize it.
- Float or Suspend: After magnetization, use the water and leaf method or suspend the needle through a piece of cork or paper, using thread or your hair.

Improvising with Natural Materials

In the absence of synthetic materials, nature offers alternatives.

- Natural Water Containers: Hollowed-out wood, large leaves, or rock indentations can hold water for the leaf and needle compass.
- Natural Magnetization: Iron-rich rocks, found in certain areas, can sometimes be used to magnetize a needle by repeatedly striking or rubbing the needle against them.

Troubleshooting and Accuracy

Improvised compasses, while ingenious, come with limitations and potential inaccuracies.

- Calibration: If possible, test your improvised compass against a known direction to calibrate it.
- Environmental Factors: Metal objects, electrical sources, and large iron deposits in the earth can affect the accuracy of a magnetic compass.
- Sun Movement: Remember, the shadow-stick method relies on the movement of the sun and can only be used during daylight hours.

Ethical Considerations and Environmental Impact

As with all survival skills, the crafting and use of improvised compasses should be approached with respect for nature and the environment.

- Minimal Disturbance: Use materials that have minimal impact on the environment. Avoid damaging living plants or altering natural landscapes.
- Leave No Trace: After using natural water sources or creating markers, restore the area to its original state as much as possible.

Mastering the craft of improvised compasses equips you with a valuable skill set for survival navigation. It underscores the importance of resourcefulness and adaptability in the wild. While these methods may not replace a traditional compass in accuracy, they offer crucial directional guidance when conventional tools are unavailable. As we venture further into survival navigation techniques, remember that the essence of navigation lies not just in the tools we use but in our connection with the natural world and our understanding of its principles. This knowledge, combined with creativity and respect for nature, ensures that we can find our way, even under the most challenging circumstances.

Chapter 3: Safe and Effective Night Navigation Strategies

Navigating the wilderness at night presents a unique set of challenges and risks, yet with the right knowledge and skills, it can also be a rewarding experience that sharpens your survival instincts. This chapter delves into strategies for safe and effective night navigation, empowering you to move confidently through the darkness while minimizing the potential for danger.

Understanding Night Navigation

Night navigation requires a heightened sense of awareness and an adaptation to relying more heavily on non-visual senses. The darkness significantly reduces visibility, making familiar landscapes seem unfamiliar and concealing hazards.

Preparing for Night Navigation

Preparation is key to successful night navigation. Before the sun sets, take the time to prepare:

- Plan Your Route: If possible, scout your route during daylight. Identify landmarks that can guide you at night.
- Gather Necessary Tools: Equip yourself with a reliable flashlight or headlamp, spare batteries, and glow sticks for marking your path or signaling for help.
- Inform Others: Always let someone know your planned route and expected return time.

Using Natural Light Sources

The moon and stars can be invaluable allies in night navigation, offering both light and directional guidance.

Moon Navigation

The moon rises in the east and sets in the west, similar to the sun. Its phase can also provide directional clues:

- Full Moon: Offers the most light, casting shadows that can help define the terrain.

- Crescent Moon: The tips point away from the sun, helping to orient you if you know the sun's approximate set or rise position.

Star Navigation

Familiarity with key constellations can serve as a compass in the night sky:

- The North Star (Polaris): In the Northern Hemisphere, locating Polaris gives you a true north direction.
- The Southern Cross: In the Southern Hemisphere, extending the long axis of the Southern Cross points south.

Enhancing Night Vision

Preserving your night vision is crucial for navigating in the dark:

- Red Light: Use a flashlight with a red filter or a red headlamp. Red light minimizes the impact on your night vision compared to white or blue light.
- Adaptation Time: Allow your eyes to adjust to the darkness for about 20-30 minutes. Avoid looking directly at bright lights to maintain night adaptation.

Listening and Feeling Your Way

Your senses of hearing and touch become more critical when visual cues are limited:

- Sound Cues: Pay attention to the sounds of running water, animal calls, or the rustling of leaves, which can provide directional cues and alert you to nearby water sources or trails.
- Ground Feedback: Feel the ground beneath your feet. Trails and paths often feel different from the surrounding terrain.

Navigating by the Stars

Basic knowledge of celestial navigation can be a powerful tool at night:

- Learn Key Constellations: Besides Polaris and the Southern Cross, familiarize yourself with other major constellations and their seasonal positions in the sky.
- Celestial Movement: Remember that stars move across the sky throughout the night. Use their movement relative to fixed landmarks to maintain your direction.

Safety Precautions

Night navigation comes with increased risks, making safety precautions paramount:

- Pace Yourself: Move more slowly than you would during the day to avoid injuries.
- Use Markers: Place reflective markers or glow sticks to mark your path and key locations.
- Stay Alert: Be aware of your surroundings and potential hazards such as cliffs, water bodies, and dense underbrush.

When to Stay Put

Sometimes, the safest strategy is not to navigate but to stay put:

- Hazardous Conditions: In unfamiliar terrain or adverse weather conditions, consider setting up a temporary camp until daylight.
- Energy Conservation: If you're exhausted, a rest can recharge your energy for safer travel at first light.

Psychological Aspects

Night navigation can be as much a psychological challenge as a physical one:

- Stay Calm: Fear of the dark is natural. Maintain a calm mindset by focusing on tasks and navigation.
- Confidence in Skills: Trust in your preparation and skills to navigate the night.

Navigating through the night is an advanced survival skill that demands respect for the challenges it presents. By preparing adequately, understanding the natural cues available, and taking necessary safety precautions, you can navigate effectively and safely under the cover of darkness. This chapter's strategies are designed to equip you with the knowledge to embrace the night confidently, turning what might seem like an insurmountable challenge into an opportunity for growth and learning in your survival skills repertoire. Remember, the key to successful night navigation lies not only in mastering technical skills but also in developing a deep connection with the natural world and trusting in your ability to adapt and overcome.

Book 7:
Wilderness First Aid

Chapter 1: Essential Elements of a Survival First Aid Kit

A survival first aid kit is a crucial companion for any wilderness adventure, embodying preparedness and the ability to respond effectively to injuries and health emergencies that may arise in the wild. Unlike standard first aid kits, a survival kit must be meticulously curated to address the unique challenges and risks of remote environments, balancing comprehensiveness with portability. This chapter outlines the essential components of a survival first aid kit, ensuring you're equipped to handle common wilderness injuries and emergencies.

Fundamental Supplies

Sterile Gauze and Bandages

- Purpose: For dressing wounds, stopping bleeding, and protecting injured areas.
- Variety: Include various sizes of adhesive bandages, gauze pads, and rolls. Non-adhesive bandages are crucial for sensitive wounds.

Antiseptic Wipes and Ointments

- Purpose: To clean wounds and prevent infection.
- Types: Antiseptic wipes for initial cleaning, and antibiotic ointment or cream for post-cleaning application.

Medical Tape and Closure Strips

- Purpose: To secure bandages and close minor lacerations.
- Selection: Waterproof medical tape and butterfly closure strips or steri-strips for wound closure without sutures.

Tweezers and Scissors

- Purpose: For removing splinters, thorns, or ticks, and cutting bandages or medical tape.
- Specifications: Fine-point tweezers for precision and safety scissors with a blunt end to prevent accidental injury.

Specialized Equipment

Splinting Materials

- Purpose: To immobilize fractures or sprains.
- Options: Flexible splints, like SAM splints, are versatile and can be molded to fit various limbs. Include elastic bandages or medical wrap for securing splints.

Blister Treatment

- Purpose: To prevent and manage blisters, a common issue in rigorous outdoor activities.
- Components: Moleskin or blister pads, and antiseptic wipes to clean the area before application.

Thermal Blanket

- Purpose: To manage body temperature, crucial in preventing hypothermia or shock.
- Feature: Compact, lightweight, and designed to retain body heat. Reflective blankets can also signal for help.

Medications

Over-the-Counter Medications

- Purpose: To manage pain, inflammation, allergies, and gastrointestinal issues.
- Essentials: Include ibuprofen or acetaminophen for pain and fever, antihistamines for allergic reactions, antidiarrheal medication, and electrolyte tablets for hydration.

Prescription Medications

- Consideration: If you or your group members require specific prescription medications, ensure an adequate supply is included in the kit.

Advanced Supplies

Hemostatic Agents

- Purpose: To rapidly stop severe bleeding.
- Examples: Hemostatic gauze or powders that can accelerate clotting.

Suture Kit or Wound Closure Strips

- Usage: For closing deep cuts when professional medical help is not immediately accessible.
- Note: Only attempt wound suturing if you have the proper training.

Personal Protection

Gloves

- Purpose: To protect both the caregiver and the patient from infection.
- Recommendation: Include multiple pairs of nitrile gloves, which are less likely to cause allergic reactions than latex.

CPR Mask

- Purpose: To safely perform CPR.
- Advantage: A barrier device like a CPR mask can make rescue breathing more effective and hygienic.

Navigation and Communication Tools

Whistle

- Purpose: For signaling help.
- Advantage: A whistle can be heard over longer distances than the human voice and requires minimal energy to use.

Waterproof Notepad and Pencil

- Purpose: To document medical care, symptoms, or vital signs.
- Importance: Keeping records can be vital for ongoing treatment, especially if rescue teams are involved.

Customizing Your Kit

While the listed items constitute the core of a survival first aid kit, customization based on the environment, group size, and specific medical needs or skills is crucial. Consider the following:

- Environment-Specific Items: Snake bite kits in snake-prone areas, water purification tablets, or sunburn relief gel.
- Skill Level: Tailor your kit to match your medical training. Advanced items are only useful if you know how to safely use them.
- Group Needs: For group excursions, adjust the quantity of supplies and include items for common ailments among members.

Assembling a survival first aid kit is a thoughtful process that reflects a commitment to safety and preparedness. By prioritizing versatility, practicality, and the specific needs of your adventure, you can create a kit that not only meets the essential criteria for wilderness survival but also embodies your readiness to face the challenges of the wild with confidence. Remember, a well-prepared first aid kit is not just about the items it contains, but about the peace of mind and the ability to respond effectively to emergencies it provides.

Chapter 2: Dealing with Common Injuries in the Wild

Venturing into the wilderness brings a sense of freedom and connection to nature, but it also exposes adventurers to the risk of injuries. From minor cuts and blisters to more severe conditions like fractures or animal bites, being prepared to handle common injuries is a fundamental aspect of wilderness survival. This chapter provides insights and practical advice on managing these injuries, ensuring that you can respond effectively when faced with medical challenges in remote settings.

Cuts and Scrapes

Cuts and scrapes are among the most frequent injuries encountered in the wild. Though often minor, they require proper care to prevent infection.

- First Aid Steps:
  - Clean the Wound: Rinse the wound with clean water to remove debris. Use antiseptic wipes or solutions to cleanse the area around the wound.
  - Stop the Bleeding: Apply direct pressure with a clean cloth or gauze until bleeding stops.
  - Protect the Wound: Cover the wound with a sterile bandage or gauze and secure it with medical tape.

Blisters

Blisters can turn a rewarding hike into a painful ordeal. Prevention is key, but once a blister forms, proper management is crucial.

- Prevention: Wear well-fitting footwear and moisture-wicking socks. Apply moleskin or blister pads to high-friction areas.
- Care: If a blister forms, cover it with a blister bandage or moleskin. If it's painful and likely to burst, sterilize a needle with alcohol, gently puncture the blister's edge, press out the fluid, and apply an antibiotic ointment and bandage.

Sprains and Strains

Ligaments (sprains) and muscles or tendons (strains) are often injured in rugged outdoor activities. The R.I.C.E. method is the cornerstone of treatment.

- R.I.C.E.: Rest, Ice, Compression, and Elevation.
  - Rest: Avoid putting weight on the injured area.
  - Ice: Apply cold packs for short periods to reduce swelling.
  - Compression: Use an elastic bandage to apply gentle pressure.
  - Elevation: Keep the injured area raised above heart level to decrease swelling.

Fractures

Fractures in the wilderness pose a significant challenge, requiring stabilization until professional medical help can be reached.

- Immobilization: Use splints made from sticks, trekking poles, or even folded items to immobilize the fractured limb. Ensure the splint extends beyond the joints above and below the fracture.
- Pain Management: Administer over-the-counter pain relievers if available and safe for the injured person.
- Evacuation: Fractures often require evacuation. Signal for help or carefully transport the injured person if it's safe to move.

Animal and Insect Bites

Wilderness areas are habitats for various animals and insects, some of which might bite or sting.

- Insect Bites and Stings: Remove stingers with a scraping motion using a credit card or fingernail. Apply a cold pack to reduce swelling and consider antihistamines for allergic reactions.
- Snake Bites: Keep the bitten limb immobilized and at heart level. Do not attempt to suck out the venom. Seek immediate medical assistance.
- Large Animal Bites: Clean the wound thoroughly but do not close it, as this can trap bacteria inside. Seek medical help as soon as possible for further treatment and rabies assessment.

Hypothermia and Frostbite

Cold environments can lead to hypothermia and frostbite, conditions that require immediate attention.

- Hypothermia: Remove any wet clothing and replace it with dry, warm layers. Share body heat or use emergency blankets to warm the person. Offer warm, sweetened liquids if they're conscious.
- Frostbite: Warm the affected areas slowly by placing them close to the body or in warm (not hot) water. Do not rub frostbitten skin.

Heat-Related Illnesses

Heat exhaustion and heatstroke are serious risks during activities in hot climates.

- Heat Exhaustion: Move to a cool place, loosen clothing, and hydrate with water or sports drinks. Apply cool, wet cloths to the skin.
- Heatstroke: This is a medical emergency. Cool the person with whatever means available while waiting for emergency services. Immersion in cool water or applying cold packs to the body's major arteries can be effective.

Burns

Burns require careful management to prevent infection and promote healing.

- First Aid for Burns: Cool the burn under running water for at least 10 minutes. Cover loosely with a sterile dressing. Do not apply ice, butter, or ointments, which can cause further damage.

Dealing with common injuries in the wild demands a combination of preparedness, knowledge, and calm decision-making. A well-stocked first aid kit is your first line of defense, but understanding how to use its contents effectively is equally important. By familiarizing yourself with these first aid techniques and principles, you equip yourself to manage injuries that may occur in the wilderness, ensuring that your outdoor adventures remain safe and enjoyable. Remember, when in doubt or faced with severe injuries, seek professional medical assistance as soon as possible. Your ability to respond promptly and effectively can make a significant difference in the outcome of wilderness injuries.

Chapter 3: Natural Remedies Using Medicinal Plants

In the heart of wilderness survival lies the ancient wisdom of using medicinal plants for healing. Beyond the contents of a first aid kit, the natural world offers a bounty of remedies that can address a wide range of ailments and injuries. This chapter delves into the art and science of utilizing medicinal plants, guiding you through identifying, harvesting, and applying nature's healing agents with respect and caution.

The Foundation of Plant-Based Healing

The use of plants for medicinal purposes is as old as humanity itself, forming the basis of traditional medicine systems worldwide. Understanding the properties of various plants and how they can be used to treat common wilderness injuries and ailments can enhance your survival skills and reduce your reliance on manufactured medicines.

Identifying Medicinal Plants

Accurate identification is crucial when using plants for medicinal purposes. Misidentification can lead to the use of toxic or harmful species.

- Field Guides and Local Knowledge: Invest in a reputable field guide to medicinal plants specific to the area you're exploring. Whenever possible, learn from local experts or indigenous peoples familiar with regional flora.
- Physical Characteristics: Pay close attention to the plant's leaves, flowers, stems, and roots—each detail can be critical for proper identification.
- Habitat: Note the environment where the plant thrives, as many species have specific habitat preferences that can aid in identification.

Common Medicinal Plants and Their Uses

Several plants are renowned for their medicinal properties and can be found in various wilderness settings.

Willow Bark (Salix spp.)

- Uses: The bark of willow trees contains salicin, a compound similar to aspirin. It can be used to reduce fever, alleviate pain, and diminish inflammation.
- Preparation: The bark can be chewed directly or steeped in hot water to make a tea.

Plantain (Plantago major)

- Uses: Not to be confused with the banana-like fruit, this common weed has potent antiseptic and anti-inflammatory properties. It's excellent for cuts, scrapes, and bug bites.
- Preparation: Crush the leaves to release the juice and apply directly to the affected area, or use them to make a poultice.

Lavender (Lavandula spp.)

- Uses: Known for its calming and antiseptic properties, lavender can be used to treat anxiety, insomnia, wounds, and burns.
- Preparation: Lavender flowers can be used to make tea, or the oil can be applied directly to the skin for its soothing effects.

Aloe Vera (Aloe barbadensis miller)

- Uses: Aloe vera gel is famous for its soothing, anti-inflammatory, and healing properties, especially for burns and skin irritations.
- Preparation: Cut a leaf open and apply the gel directly to the affected area.

Chamomile (Matricaria chamomilla)

- Uses: Chamomile is widely used for its calming effects and can help with sleep disturbances, anxiety, and digestive issues.
- Preparation: The flowers can be dried and used to make a soothing tea.

Harvesting and Preparation

Harvesting plants for medicinal use requires respect for the environment and knowledge of sustainable practices.

- Sustainable Harvesting: Only take what you need, and never harvest more than a small percentage of plants from any given area to ensure populations remain healthy.

- Ethical Considerations: Be mindful of the ecological role of the plants you harvest and the impact your actions may have on local wildlife and plant communities.

Safety and Considerations

While medicinal plants can offer significant benefits, they also come with cautions.

- Allergic Reactions: Test a small amount of any new remedy on a patch of skin before full use to check for allergic reactions.
- Pregnancy and Medical Conditions: Certain plants can be harmful during pregnancy or may interact with existing medical conditions and medications.
- Consultation: Whenever possible, consult with a healthcare professional or a knowledgeable herbalist before using medicinal plants, especially for serious conditions.

Integrating Medicinal Plants into Wilderness First Aid

Incorporating plant-based remedies into your wilderness first aid approach adds a valuable dimension to your ability to care for yourself and others.

- Complementary Use: Use medicinal plants in conjunction with conventional first aid treatments where appropriate.
- Knowledge Sharing: Teach fellow adventurers and group members about the safe and effective use of medicinal plants.

Embracing the use of medicinal plants in wilderness survival not only connects us with ancient healing traditions but also deepens our relationship with the natural world. As we learn to identify and utilize the healing powers of plants, we gain invaluable allies in maintaining health and treating injuries in remote settings. However, this knowledge comes with the responsibility to harvest and use these resources ethically and safely, always prioritizing the well-being of both the individual and the environment. Armed with understanding and respect, we can harness the healing power of the earth, ensuring that our adventures in the wild are safer, more sustainable, and deeply enriched by the natural remedies that surround us.

Chapter 4: Managing Environmental Hazards Like Hypothermia and Heatstroke

The wilderness offers a retreat into the beauty and serenity of nature, but it also presents a unique set of environmental hazards. Among these, hypothermia and heatstroke are two extremes on the temperature spectrum that can pose significant risks to outdoor adventurers. Understanding how to manage these conditions is crucial for anyone venturing into the wild. This chapter focuses on the identification, prevention, and treatment of hypothermia and heatstroke, empowering you with the knowledge to handle these environmental challenges safely.

Understanding Hypothermia

Hypothermia occurs when the body loses heat faster than it can produce it, causing the core body temperature to drop below 95°F (35°C). It can happen in any environment, not just in cold climates, especially when an individual is wet, tired, and exposed to wind or cold water.

Signs and Symptoms

- Shivering, which may stop as hypothermia progresses
- Slurred speech or mumbling
- Slow, shallow breathing
- Weak pulse
- Clumsiness or lack of coordination
- Drowsiness or very low energy
- Confusion or memory loss

Prevention

- Wear appropriate clothing: Layering is key. Use moisture-wicking fabrics next to the skin and waterproof, windproof layers on the outside.
- Stay dry: Keep dry with waterproof gear, and change out of wet clothes as soon as possible.
- Eat and drink: Consume high-energy foods and stay hydrated to help your body produce heat.
- Stay active: Keep moving to generate body heat, but avoid sweating.

Treatment

- Move the person to a warm, sheltered area.

- Remove wet clothing and replace it with dry, warm layers.
- Insulate the individual from the cold ground.
- Share body heat if necessary.
- Offer warm, sweetened drinks if the person is conscious.
- Seek medical help immediately.

Understanding Heatstroke

Heatstroke is the most severe form of heat illness, occurring when the body overheats and cannot cool down effectively, usually as a result of prolonged exposure to high temperatures or physical exertion in hot conditions. This condition can be life-threatening and requires immediate action.

Signs and Symptoms

- High body temperature (104°F or 40°C or higher)
- Altered mental state or behavior (confusion, agitation, slurred speech)
- Nausea and vomiting
- Flushed skin
- Rapid, shallow breathing
- Racing heart rate
- Headache

Prevention

- Hydrate: Drink plenty of fluids before, during, and after exposure to heat.
- Wear appropriate clothing: Choose lightweight, loose-fitting, and light-colored clothes.
- Acclimatize: Gradually get used to the heat over several days.
- Seek shade: Use hats, sunglasses, and find shelter to avoid direct sun exposure.
- Rest: Take regular breaks in a cool area.

Treatment

- Move the person to a cooler place immediately.
- Remove excess clothing to expose as much skin as possible.
- Cool the person rapidly using whatever means available: immerse in cool water, place in a cold shower, use wet towels, or fan while misting with water.

- Offer cool water to drink if the person is conscious and able to swallow.
- Seek emergency medical help without delay.

Handling Both Extremes

Managing environmental hazards like hypothermia and heatstroke involves a delicate balance of preparation, awareness, and quick response. Here are additional tips that apply to handling both conditions:

- Education: Know the signs and symptoms of both conditions to recognize them early.
- Monitoring: Keep a close eye on yourself and your group, especially those more vulnerable to temperature-related illnesses.
- Communication: Have a means of communication for emergency situations, whether it's a whistle, a mirror for signaling, or a fully charged phone for areas with coverage.
- Emergency Kit: Always carry an emergency kit with thermal blankets, additional water, sunscreen, and high-energy snacks.

Venturing into the wilderness requires respect for nature's power and an understanding of how to coexist safely with the elements. Hypothermia and heatstroke represent significant risks, but with proper preparation, knowledge, and vigilance, these conditions can be effectively managed and even prevented. This chapter has equipped you with the critical information needed to recognize, prevent, and treat these environmental hazards, ensuring your wilderness adventures remain safe and enjoyable. Remember, the best way to combat the extremes of cold and heat is through education, preparation, and a commitment to safety.

Book 8:
Weather and Environment Adaptation

Chapter 1: Predicting Weather Changes and Preparing Accordingly

The ability to predict and prepare for weather changes is not just a survival skill—it's an essential part of interacting safely and respectfully with the natural world. This chapter delves into the basics of weather prediction using both traditional and modern methods, providing you with the knowledge to read the signs of impending weather shifts and take appropriate action to safeguard yourself and your companions in the wilderness.

Understanding Weather Patterns

Weather is the state of the atmosphere at any given time and place, influenced by factors such as temperature, humidity, air pressure, wind, and precipitation. Learning to recognize the signs of changing weather patterns is crucial for planning your activities and ensuring safety in the outdoors.

Cloud Types and What They Indicate

- Cirrus Clouds: These high, wispy clouds often indicate fair weather, but when they thicken and blanket the sky, they can signal an approaching warm front, possibly leading to precipitation within the next 24 to 36 hours.
- Cumulus Clouds: Fluffy and white with flat bases, cumulus clouds suggest pleasant weather. However, if they grow taller and darker (cumulonimbus), they warn of thunderstorms.
- Stratus Clouds: These low, gray clouds cover the sky like a blanket, often bringing light rain or drizzle. If they descend even lower, fog is likely.

Barometric Pressure Changes

A barometer measures air pressure. Falling pressure typically indicates an approaching storm, while rising pressure suggests improving weather. Many outdoor watches and portable weather stations include barometers.

Wind Direction Shifts

Wind direction can offer clues about weather changes. For example, in many regions, a shift to a wind from the east often precedes bad weather, while a shift to a wind from the west indicates fair weather.

Traditional Weather Prediction Methods

Long before the advent of modern meteorology, people relied on observations of nature to predict the weather. These traditional methods can still be useful, especially in remote areas.

Animal Behavior

Many animals are sensitive to changes in air pressure and humidity and will alter their behavior accordingly:

- Birds: Birds flying high in the sky often indicate fair weather, while flying low can mean bad weather is coming, as birds avoid flying at high altitudes during low air pressure.
- Insects: Increased mosquito activity and the appearance of spiders indoors can indicate an approaching storm.

Plant Responses

Some plants respond visibly to changes in humidity and pressure:

- Pine Cones: Pine cones open in dry air and close in humid air, indicating an increase in moisture and possibly rain.
- Flowers: Certain flowers, like tulips and daisies, close their petals when bad weather is approaching.

Preparing for Weather Changes

Clothing and Gear

- Layering: Dress in layers that can be easily added or removed to adjust to changing temperatures.

- Protection: Always carry waterproof and windproof gear, even if the forecast is clear. Conditions can change rapidly, especially in mountainous areas.

Shelter and Campsite Selection

- High Ground: Avoid setting up camp in low-lying areas prone to flooding during rain.
- Natural Shelter: Use natural features like rock overhangs or dense forests to provide additional protection from elements.

Emergency Plans

- Communication: Carry a means of communication, such as a satellite phone or a personal locator beacon, especially in remote areas.
- Emergency Kit: Include a whistle, thermal blanket, flashlight, extra food, and water in your emergency kit, tailored to address both the expected weather and potential unexpected changes.

Modern Tools and Resources

While traditional methods provide valuable insights, modern tools and resources offer precision and reliability in weather prediction.

Apps and Online Resources

Numerous weather apps and websites provide up-to-date forecasts and weather alerts. Apps like NOAA Weather Radar and The Weather Channel offer detailed information, including satellite imagery and radar maps.

Portable Weather Stations

Portable weather stations can measure temperature, humidity, barometric pressure, and wind speed, giving you real-time data to make informed decisions about your plans.

Predicting weather changes and preparing accordingly is a blend of art and science, requiring attention to both the subtle cues provided by nature and the detailed information available through modern technology. By developing your skills in weather observation and learning to interpret the signs, you can enhance your ability to adapt to the environment, ensuring not only your safety but also a deeper connection with the natural world around you. Whether you're planning a day hike or a prolonged wilderness expedition, being weather-wise is an indispensable part of outdoor preparedness, empowering you to face the challenges and joys of the wilderness with confidence and respect.

Chapter 2: Survival Strategies for Extreme Cold Conditions

Surviving in extreme cold conditions challenges both the mind and body, demanding preparation, knowledge, and resilience. The stark beauty of icy landscapes can quickly become perilous without the right strategies to stay warm, hydrated, and safe. This chapter explores essential survival techniques for enduring and thriving in the biting cold, from the icy reaches of the Arctic to the frostbitten trails of high-altitude mountains.

Understanding Cold Weather Risks

Before delving into survival strategies, it's crucial to recognize the risks posed by extreme cold:

- Hypothermia: The most significant danger, where the body loses heat faster than it can produce, leading to a dangerously low body temperature.
- Frostbite: The freezing of skin and underlying tissues, usually affecting extremities like fingers, toes, ears, and nose.
- Dehydration: Cold air contains less moisture, and the body's thirst response is diminished, increasing the risk of dehydration.
- Snow Blindness: Overexposure to UV rays reflected off snow can cause temporary vision loss.

Layered Clothing: The First Defense

The cornerstone of cold weather survival is proper clothing, arranged in layers for flexibility and efficiency:

- Base Layer: Moisture-wicking materials like merino wool or synthetic fabrics keep the skin dry by moving sweat away.
- Insulating Layer: Wool, fleece, or down provide warmth by trapping air close to the body.
- Outer Layer: A windproof and waterproof shell protects against wind, snow, and rain.

Shelter: Your Refuge from the Cold

In extreme cold conditions, a well-constructed shelter is vital for survival, providing protection from the elements and helping to conserve body heat.

- Natural Shelters: Snow caves or igloos can be effective but require knowledge and effort to construct safely. The insulation properties of snow can help maintain a warmer internal temperature.
- Tent and Bivy Sacks: For more mobile survival, high-quality, four-season tents and insulated bivy sacks offer lightweight, reliable shelter options.
- Insulation from the Ground: Use branches, leaves, or a sleeping pad to insulate your body from the cold ground, which can sap heat rapidly.

Fire: The Lifeline in the Cold

Fire serves multiple roles in cold weather survival: warmth, melting snow for water, cooking, and signaling for rescue.

- Preparation: Gather a substantial stockpile of dry wood and tinder before dark. Store some inside your shelter to keep it dry overnight.
- Ignition: Use waterproof matches, lighters, or fire starters. In wet conditions, resinous wood from pine trees can help ignite and sustain a fire.
- Safety: Build your fire downwind of your shelter to prevent smoke inhalation and fire hazards.

Hydration and Nutrition

The body expends more energy in cold environments, making hydration and calorie intake crucial.

- Melting Snow: Always melt snow before drinking to prevent lowering your body temperature. Use a fire or portable stove.
- High-Energy Foods: Consume foods high in fats and proteins for sustained energy. Nuts, chocolate, and energy bars are excellent choices.

Navigating Cold Weather Hazards

- Avalanche Awareness: Learn to recognize avalanche-prone areas and carry appropriate safety gear, such as a beacon, probe, and shovel.
- Ice Safety: Test ice thickness before crossing frozen bodies of water. Carry ice picks for self-rescue in case of a fall through ice.

Cold Weather Health Monitoring

Keeping an eye on your physical condition and that of your group is essential to prevent cold-related illnesses.

- Regular Checks: Monitor for signs of frostbite and hypothermia, including uncontrolled shivering, slurred speech, and discolored skin.
- Buddy System: Use the buddy system to help monitor and assist each other with warmth and shelter.

Mental Resilience in the Cold

The psychological challenge of surviving in extreme cold cannot be underestimated. Mental resilience can be as crucial as physical preparation.

- Stay Positive: Maintain a positive attitude and focus on tasks to keep your mind occupied.
- Routine: Establish a routine to provide structure and a sense of normalcy.
- Goal Setting: Set small, achievable goals to help maintain focus and motivation.

Adaptation Through Knowledge and Skills

Survival in extreme cold conditions is not solely about battling the elements but adapting to them. Acquiring knowledge and skills before venturing into cold environments can significantly increase your chances of survival and recovery.

- Training: Participate in cold weather survival courses to gain practical experience.
- Research: Study accounts of survival in similar conditions for insights and strategies.

Surviving in extreme cold conditions demands respect for the environment and a comprehensive approach that includes proper gear, shelter, fire-making, nutrition, and hydration strategies. Equally important is the awareness of potential hazards and the cultivation of mental resilience. By preparing meticulously and applying the survival techniques outlined in this chapter, adventurers can navigate the challenges of the cold, embracing the beauty and majesty of winter landscapes with confidence and safety.

Chapter 3: Life-Saving Techniques for Arid Climate Survival

Surviving in arid climates poses unique challenges, with extreme heat, scarce water, and exposure to the sun being the primary concerns. The desert's beauty is matched by its harshness, demanding specific survival strategies to navigate its dangers effectively. This chapter focuses on life-saving techniques tailored for arid climate survival, equipping you with the knowledge to withstand the desert's extreme conditions.

Understanding the Arid Climate

Arid climates are characterized by low humidity, scarce rainfall, and high temperatures during the day with a significant drop at night. These conditions require a survival strategy that prioritizes water conservation, heat illness prevention, and effective use of available resources.

Water: The Essence of Desert Survival

Finding and conserving water is paramount in arid environments.

- Locating Water: Look for signs of water, such as animal tracks converging or vegetation patches. Dry riverbeds can contain water just below the surface. Use rocks or vegetation to collect morning dew.
- Conservation: Minimize sweating by limiting physical exertion during the hottest parts of the day. Wear light, loose-fitting, and light-colored clothing to reflect sunlight.
- Purification: Always purify found water using boiling, tablets, or a portable filter to avoid waterborne diseases.

Managing Exposure to Heat

Protecting yourself from the sun and managing the risk of heat-related illnesses are critical.

- Shade: Use natural features for shade during the peak sun hours. If natural shade is unavailable, create a makeshift shelter using clothing, blankets, or any available material.
- Clothing: Long sleeves, wide-brimmed hats, and sunglasses can protect the skin and eyes from UV damage. Clothing should be breathable to allow for ventilation.
- Sunscreen: Apply sunscreen to exposed skin areas to prevent sunburn, which can exacerbate dehydration and heat-related illnesses.

Navigating and Moving Efficiently

Effective navigation is crucial to finding resources and avoiding unnecessary energy expenditure.

- Travel Times: Plan movement during cooler times of the day, such as early morning or late afternoon. Rest in the shade during peak heat hours.
- Landmarks: Use distinctive landscape features for navigation. Arid environments often lack obvious landmarks, so pay attention to subtle terrain changes.
- Conserving Energy: Walk at a steady, slow pace to conserve energy and water. Avoid steep dunes or hills when possible.

Shelter: A Refuge from the Elements

A shelter can provide a respite from the sun and help retain warmth during cold desert nights.

- Location: Choose a shelter location that offers protection from the sun and wind. If possible, position near a potential water source.
- Insulation: Use available materials to insulate the shelter from the hot ground and retain warmth at night. Sands can be an insulator; digging slightly below the surface can reveal cooler layers.
- Ventilation: Ensure your shelter has ventilation to prevent overheating during the day.

Fire: A Tool for Warmth and Signaling

While fire is less critical for warmth in the desert, it remains essential for cooking, boiling water, and signaling for rescue.

- Fuel: Collect dry vegetation, wood, or any combustible material during your travels. Desert plants can be surprisingly flammable.
- Use Sparingly: Given the scarcity of fuel, use fire judiciously. A small, controlled fire is usually sufficient for cooking or boiling water.

Signaling for Rescue

Being visible to rescuers is vital in expansive desert terrains.

- Mirrors and Reflective Objects: Use mirrors or any reflective material to signal aircraft. The sun's reflection can be seen for miles.
- Smoke Signals: If you have enough materials, create a smoky fire for signaling. Green vegetation produces more smoke.
- Marking the Ground: Use rocks or sand to create large, visible distress signals or arrows pointing to your location.

Adapting to Temperature Variations

Desert temperatures can plummet at night, requiring strategies to stay warm.

- Layering: Utilize layers of clothing to adjust to changing temperatures. Pack a lightweight, insulated jacket even if daytime temperatures are high.
- Stay Dry: Damp clothing can lead to significant heat loss at night. Change into dry clothes before sunset if possible.

Encountering Wildlife

Deserts are home to a variety of wildlife, some of which can be dangerous.

- Avoidance: Keep food sealed and avoid areas where dangerous animals may congregate, such as waterholes at night.
- Awareness: Be aware of your surroundings, especially when moving through vegetation or rocky areas where snakes or scorpions may be hidden.

Survival in arid climates demands respect for the environment and an understanding of its extremes. By prioritizing water sourcing and conservation, protecting yourself from the sun and heat, and utilizing effective shelter and navigation strategies, you can enhance your chances of not only surviving but thriving in the desert. Adaptability, preparation, and a keen awareness of your surroundings are your best allies in the stark, beautiful, and challenging desert landscape.

Chapter 4: Adapting to Wetland and Rainforest Environments

Wetlands and rainforests are ecosystems rich in biodiversity and natural beauty. However, they also present unique challenges for survival, including high humidity, frequent rainfall, dense vegetation, and a diverse array of wildlife. Adapting to these environments requires specialized knowledge and preparation. This chapter explores strategies for survival and adaptation in the lush but challenging terrains of wetlands and rainforests.

Understanding the Environment

Wetlands and rainforests are characterized by their high moisture levels, either standing water or frequent precipitation, leading to high humidity. These conditions can affect personal comfort, health, and the ability to move or find resources.

Key Challenges:

- Navigation Difficulties: Dense foliage can limit visibility, making navigation challenging.
- Water Availability: Despite abundant water, finding safe drinking sources can be difficult due to contamination.
- Pests and Wildlife: These areas are home to insects and animals that can pose threats to humans.
- Rot and Dampness: High moisture levels can lead to rapid deterioration of equipment and clothing.

Survival Strategies

Water Procurement and Purification

- Rainwater Collection: Use broad leaves or tarps to collect rainwater, a reliable and relatively clean water source.
- Water Purification: Always purify collected water using boiling, filters, or chemical treatment methods to avoid waterborne diseases.

Shelter Building

- Elevated Structures: In wetlands, construct shelters off the ground to avoid water and dampness. Use stilts or find naturally elevated areas.
- Waterproofing: In rainforests, ensure your shelter has a waterproof roof made from broad leaves or tarps. Angling the roof can help direct rainwater away from your sleeping area.

Clothing and Gear

- Moisture-Wicking Fabrics: Choose clothing that wicks moisture away from the body to help manage sweat and prevent chafing.
- Waterproof Gear: Waterproof backpacks and containers protect essential items from moisture. Sealable plastic bags can offer additional protection for critical items like matches and electronics.

Dealing with Pests and Wildlife

- Insect Protection: Use insect repellent liberally. Wear long sleeves and pants, and consider a mosquito net for sleeping.
- Avoiding Wildlife Encounters: Store food securely and away from your campsite to avoid attracting animals. Be aware of your surroundings and make noise to alert wildlife of your presence, reducing the chance of surprise encounters.

Navigation Tips

- Use Natural Landmarks: Identify unique trees, rock formations, or waterways as reference points to aid navigation.
- Compass and GPS: Reliable navigation tools are indispensable in these environments. Learn to use them effectively in conjunction with natural navigation techniques.
- Pacing and Time: Keep track of your pacing and the time spent traveling to estimate distances covered, considering the slow movement through dense vegetation.

Health and Safety

- Avoiding Illness: Be cautious of plants and water sources that may cause illness. Familiarize yourself with local flora and fauna to identify potential hazards.
- First Aid Preparedness: Be prepared to deal with cuts, insect bites, and other minor injuries. Keep a well-stocked first aid kit and know how to use it.

Food Sources

- Edible Plants: Many wetland and rainforest environments are rich in edible plants. However, proper identification is crucial to avoid poisonous species.

- Hunting and Fishing: These environments can offer opportunities for hunting and fishing. Use snares, traps, or fishing gear, respecting local regulations and wildlife conservation efforts.

Adapting to the Climate

- Managing Humidity: High humidity can lead to exhaustion and dehydration. Take regular breaks, stay hydrated, and manage your exertion levels.
- Temperature Fluctuations: Be prepared for cooler temperatures at night, even in tropical environments. Have suitable insulation to maintain body temperature.

Environmental Consideration

- Leave No Trace: Practice minimal impact camping and travel to preserve these unique ecosystems for future generations.
- Respect Wildlife: Observe wildlife from a distance and do not disturb their natural behaviors or habitats.

Surviving and adapting to wetland and rainforest environments demand respect for the intricacies of these ecosystems. By understanding the specific challenges they pose and employing strategies to mitigate risks, you can safely explore and appreciate the rich biodiversity and beauty they offer. Preparation, awareness, and adaptability are your best tools for navigating these lush but demanding landscapes, ensuring a harmonious and sustainable interaction with some of the planet's most vibrant ecosystems.

Book 9:
Long-Term Food Preservation

Chapter 1: Traditional Meat Preservation with Smoking and Curing

In a world where the convenience of refrigeration and freezing has become almost universal, the art of traditional meat preservation through smoking and curing retains its significance. These age-old techniques not only extend the shelf life of meats but also enhance flavors, contributing to the culinary heritage of cultures worldwide. This chapter delves into the methods of smoking and curing, offering insights into how these processes work to preserve meat and how you can apply them in your own long-term food preservation efforts.

Understanding Meat Preservation

Preserving meat involves slowing down the processes that lead to spoilage. Microbial activity, oxidation, and moisture loss are the main culprits behind meat deterioration. Smoking and curing tackle these issues by creating environments that are inhospitable to bacteria, thereby extending the meat's edible life.

The Science Behind Smoking and Curing

- Curing: Curing meat involves the application of salt, nitrates, nitrites, or sugar, either through dry rubbing or in a brine solution. Salt is hygroscopic, meaning it draws moisture out of the meat (and any bacteria present) through osmosis, effectively dehydrating the meat and inhibiting bacterial growth. Nitrates and nitrites, often found in curing salts, further prevent spoilage by blocking the growth of botulism-causing bacteria and giving cured meats their characteristic pink color.
- Smoking: Smoking meat involves exposing it to smoke from burning or smoldering materials, typically wood. This process can be hot or cold. Hot smoking cooks the meat, while cold smoking does not. Smoke contains compounds that are both antioxidant (slowing oxidation and rancidity) and antimicrobial, contributing to preservation. The smoke also imparts distinctive flavors.

Preparing Meat for Smoking and Curing

Before smoking or curing, meat must be properly prepared to ensure safety and effectiveness.

- Selecting Meat: Freshness is crucial. Begin with high-quality, fresh meat, preferably from a trusted source.
- Trimming: Remove excess fat and sinew, which can affect the curing process and smoke penetration.
- Cutting: Larger cuts can be cured and smoked, but size and thickness will impact curing times and smoking duration.

Curing Meat

Curing can be done through dry rubs or wet brines. The choice depends on personal preference, the type of meat, and the desired outcome.

Dry Curing

- Process: Coat the meat thoroughly with a mixture of salt, curing salt, and other seasonings. Place it in a container to catch any drippings and store it in a cool, dry place.
- Duration: Depending on the meat's thickness, dry curing can take from a few days to several weeks.

Wet Curing (Brining)

- Process: Submerge the meat in a brine solution made of water, salt, curing salt, sugar, and desired seasonings. Keep it refrigerated or in a cool environment.
- Duration: Wet curing times vary but generally take less time than dry curing due to the brine's ability to penetrate the meat more efficiently.

Smoking Meat

Smoking can be classified into two primary methods: hot smoking and cold smoking. Each method imparts a different flavor and level of preservation.

Hot Smoking

- Temperature: Hot smoking is done at temperatures that cook the meat, typically between 190°F to 300°F (88°C to 149°C).
- Process: After curing, the meat is smoked until it reaches a safe internal temperature. The cooking process helps to further preserve the meat.
- Flavor and Preservation: Hot smoked meat has a rich, smoky flavor and is partially preserved, though it should still be consumed within a relatively short period or refrigerated.

Cold Smoking

- Temperature: Cold smoking is performed at lower temperatures, usually below 90°F (32°C), which does not cook the meat.
- Process: Cured meat is exposed to smoke for a prolonged period, from hours to days, to infuse flavor without raising the meat's temperature.
- Flavor and Preservation: Cold smoking imparts a smoky flavor and additional preservation qualities. However, because the meat is not cooked, it must be cured thoroughly to ensure safety.

Safety Considerations

- Temperature Control: Maintaining proper temperatures during smoking is crucial to prevent the growth of harmful bacteria.
- Curing Salts: Use curing salts (containing nitrates and nitrites) with caution, following recommended guidelines and proportions to avoid health risks.

Conclusion

Traditional meat preservation through smoking and curing is a testament to human ingenuity in food preservation. These methods, developed long before the advent of modern refrigeration, remain valued not only for their practical benefits in extending meat's shelf life but also for the unique flavors they impart. By understanding and applying the principles of smoking and curing, you can preserve meat safely and enjoyably, tapping into a rich tradition that connects

Chapter 2: Drying and Dehydrating a Variety of Foods

Drying and dehydrating are among the oldest food preservation techniques known to humanity, dating back thousands of years. These methods, by reducing the moisture content in foods, significantly inhibit the growth of bacteria, yeasts, and molds which require water to thrive. This chapter explores the time-honored techniques of drying and dehydrating a wide array of foods, from fruits and vegetables to meats and herbs, providing a comprehensive guide to extending the shelf life of your harvests and culinary creations.

The Principles of Drying and Dehydrating

Drying and dehydrating are based on the simple principle of removing water from food. This can be achieved through various methods, including air drying, sun drying, oven drying, and using a dedicated food dehydrator. Each method has its nuances, but the core objective remains the same: to create a food product that's less prone to spoilage, lighter in weight, and often concentrated in flavor.

The Science Behind the Process

The science of drying food is rooted in the concept of water activity. Microorganisms that cause food spoilage and decay thrive in environments with high water activity. By reducing the moisture content in food, you effectively lower its water activity, making it less hospitable to these microorganisms. Furthermore, enzymatic reactions, which can also lead to food spoilage, are slowed down in the absence of sufficient moisture.

Techniques for Drying and Dehydrating Foods

Air Drying

- Suitable for: Herbs and leafy greens.
- Method: Tie small bunches of herbs or greens and hang them in a warm, dry, well-ventilated area away from direct sunlight. Alternatively, lay them out on a clean surface or rack, ensuring good air circulation around each piece.

Sun Drying

- Suitable for: Fruits, vegetables, and meats in hot, dry climates.

- Method: Prepare foods by slicing them thinly for uniform drying. Place them on drying racks or screens, ensuring they're not touching. Cover with a net or cheesecloth to protect from insects. Place in direct sunlight, turning occasionally, and bring in at night to prevent dew from adding moisture back.

Oven Drying

- Suitable for: A broad range of foods when outdoor drying isn't feasible.
- Method: Preheat the oven to its lowest setting (usually between 140°F to 170°F). Arrange prepared food on baking sheets in a single layer. Keep the oven door slightly ajar to allow moisture to escape and use a fan to circulate air if possible.

Using a Food Dehydrator

- Suitable for: Almost all types of food, offering the most controlled and efficient method.
- Method: Slice foods thinly and evenly. Arrange on dehydrator trays without overlapping. Set the dehydrator to the recommended temperature for the specific food type, and let it run until the food is adequately dried, following the manufacturer's guidelines or reliable recipes.

Preparing Foods for Drying

- Cleaning: Ensure all food items are clean and free from debris.
- Slicing: Thin, uniform slices dry more evenly and quickly.
- Blanching: Some vegetables may require blanching to halt enzymatic action that can cause spoilage.
- Pre-treating: Certain fruits benefit from pre-treatment, such as dipping in lemon juice or ascorbic acid solutions, to preserve color and nutrient content.

Storage and Usage

Proper storage is critical to maintaining the quality of dried foods. Once completely dried, foods should be cooled to room temperature and then stored in airtight containers. Label containers with the date of drying. Dried foods can be rehydrated for cooking or eaten as is for snacks. They provide not only a nutritious food source but also a taste of the harvest year-round.

Safety Considerations

While drying and dehydrating are relatively safe preservation methods, it's important to dry foods thoroughly to prevent mold growth. It's also crucial to start with fresh, high-quality produce and meats, as drying does not improve the quality of food.

Drying and dehydrating offer versatile and effective means of preserving a wide variety of foods, allowing us to tap into the abundance of different seasons and enjoy their flavors long after their peak. These methods, deeply rooted in our culinary heritage, not only provide practical benefits in terms of food preservation but also invite creativity in the kitchen. By mastering the art of drying and dehydrating, you can ensure a diverse and nutritious diet, minimize food waste, and step into a sustainable cycle of consumption that honors the bounty of the natural world.

Chapter 3: Fermenting and Pickling for Longevity

In the vast landscape of food preservation methods, fermenting and pickling stand out for their unique blend of science, art, and tradition. These techniques not only extend the shelf life of foods but also enhance nutritional value and flavor. This chapter delves into the ancient practices of fermenting and pickling, providing a guide to harnessing these methods for long-term food preservation.

The Art and Science of Fermentation

Fermentation is a metabolic process that converts sugar to acids, gases, or alcohol in the absence of oxygen. It's driven by beneficial microorganisms, such as bacteria and yeast, which thrive under controlled conditions to preserve and transform food.

Key Benefits

- Preservation: Fermentation produces organic acids and alcohol, natural preservatives that inhibit the growth of harmful bacteria.
- Enhanced Nutrition: Many fermented foods are richer in vitamins and probiotics, aiding digestion and improving gut health.
- Flavor Development: Fermentation can introduce complex, rich flavors to otherwise bland or unremarkable foods.

Popular Fermented Foods

- Sauerkraut: Fermented cabbage, known for its tangy flavor and digestive benefits.
- Kimchi: A Korean staple, made from fermented vegetables with a mix of seasonings.
- Yogurt: Fermented milk, thickened and imbued with beneficial bacteria.
- Kefir: A fermented milk drink, similar to yogurt but with a thinner consistency.

The Process of Pickling

Pickling involves preserving foods in an acidic solution, usually vinegar, or through the natural fermentation process where lactic acid is produced. While pickling doesn't always involve fermentation, lacto-fermented pickles are a category within this preservation method.

Vinegar Pickling

- Vinegar: Acts as a preservative due to its acidic nature, which many harmful bacteria cannot survive in.
- Flavorings: Spices and herbs are often added to the pickling solution to enhance flavor.
- Quick Pickles: A fast method where vegetables are covered in vinegar and spices, refrigerated, and consumed within a short time frame.

Lacto-Fermented Pickling

- Salt: Used to draw moisture out of the food, creating a brine that supports the growth of lactic acid bacteria.
- Anaerobic Environment: Requires that the food is submerged under the brine to prevent exposure to air, which could lead to spoilage.
- Probiotic-Rich: Lacto-fermented pickles are noted for their probiotic content, contributing to gut health.

Getting Started with Fermenting and Pickling

The process of fermenting and pickling begins with choosing high-quality, fresh ingredients. Cleanliness is crucial throughout the process to ensure that only beneficial microorganisms are introduced to your foods.

Basic Equipment

- Jars and Containers: Glass jars with airtight lids are ideal for small-batch fermentation and pickling. For larger quantities, ceramic crocks can be used.
- Weights: Keeping the food submerged under the brine is essential; food-grade weights or smaller jars can be used for this purpose.
- Cloths and Bands: To cover jars during fermentation, allowing gases to escape while keeping contaminants out.

Safety Tips

- Monitor for Signs of Spoilage: Mold, unpleasant odors, or discolored brine can indicate spoilage. If in doubt, it's safer to discard the batch.
- Use the Correct Salt Concentration: Too little salt can fail to inhibit harmful bacteria, while too much can prevent fermentation.
- pH Levels: For pickling, the pH should be 4.6 or lower to ensure safety. Test strips can be used to monitor acidity.

Creative Possibilities

Fermenting and pickling open a world of culinary creativity, allowing you to experiment with flavors, textures, and ingredients. From traditional recipes to innovative combinations, these methods invite exploration and personalization.

Storing Fermented and Pickled Foods

Proper storage is key to maintaining the longevity of fermented and pickled foods.

- Refrigeration: Slows fermentation and preserves the flavor and texture of the food.
- Canning: Some pickled foods can be canned for longer shelf life, though this process can diminish probiotic content.

Fermenting and pickling are not just preservation methods; they are a connection to our culinary heritage and a testament to the ingenuity of our ancestors. These techniques offer a sustainable approach to food preservation, reducing waste and enhancing our diets with flavorsome, nutritious foods. By embracing the practices of fermenting and pickling, we can enjoy the bounty of the harvest throughout the year, benefiting from the added health advantages and the profound satisfaction of creating something truly unique from the natural abundance around us.

Chapter 4: Natural Underground Storage Techniques

The ingenious use of the earth itself for food preservation is a practice as ancient as agriculture. Before the advent of modern refrigeration, our ancestors discovered that the stable temperatures and humidity levels found below the earth's surface could be utilized to store food for extended periods. This chapter explores the traditional and practical methods of natural underground storage, offering insights into leveraging the earth's natural properties for long-term food preservation.

The Basics of Underground Storage

Underground storage capitalizes on the earth's ability to maintain consistent temperatures and humidity levels year-round. Just a few feet below the surface, the temperature remains relatively constant, insulated from the seasonal fluctuations experienced above ground. This natural stability makes underground storage ideal for a variety of foods, including root vegetables, tubers, fruits, and even some types of meat and dairy products.

Types of Underground Storage

- Root Cellars: The most well-known form of underground storage, root cellars are buried spaces that can range from simple dugouts to more sophisticated structures with ventilation, shelving, and moisture control systems.
- Clamp Storage: A simpler method involving the storage of vegetables like potatoes and carrots in mounded piles covered with straw and earth. This method is less labor-intensive to create but offers less control over environmental conditions.
- Earth Pits: Deeper holes or pits dug into the ground, lined with straw or leaves, filled with produce, and then covered with earth. This method is particularly useful in colder climates where the ground provides additional frost protection.

Constructing a Root Cellar

Building a root cellar involves careful planning and consideration of location, drainage, ventilation, and access. A well-constructed root cellar can provide an ideal environment for preserving a wide array of produce through the cold months and beyond.

- Location: Choose a site with good drainage to prevent water accumulation. A slope can facilitate natural drainage and make excavation easier.
- Ventilation: Proper air circulation is crucial to prevent the buildup of ethylene gas and to maintain the right balance of humidity and temperature. Vents or pipes can be installed to achieve this.
- Insulation: While the earth provides natural insulation, additional insulation may be necessary in colder climates to prevent freezing.
- Accessibility: Consider how you will access your root cellar. An external entrance can reduce heat loss from the home, but an internal entrance may be more convenient.

Preparing and Storing Food

Not all foods are suited for underground storage, and different foods have different requirements for temperature, humidity, and ventilation.

- Vegetables: Root vegetables such as potatoes, carrots, and beets are ideal for underground storage. They should be cleaned gently to remove soil but not washed, as moisture can promote decay.
- Fruits: Apples, pears, and other hardy fruits can be stored, but they should be kept separate from vegetables to prevent ethylene gas from accelerating spoilage.
- Packaging: Store foods in breathable containers such as baskets, wooden crates, or burlap sacks to allow for air circulation.

Managing Temperature and Humidity

The key to successful underground storage is managing the internal environment of the storage space.

- Temperature: Ideal temperatures vary by food but generally range from 32°F to 40°F (0°C to 4.5°C) for most root vegetables and fruits.
- Humidity: A humidity level of 85% to 95% is optimal for most produce to prevent drying out. Stones, clay, or damp sand on the floor can help maintain humidity levels.

Pest Control

Underground storage areas can attract pests and rodents. Implementing measures to deter these unwanted guests is essential for the safety and integrity of stored food.

- Physical Barriers: Ensure the cellar or storage area is well sealed to prevent entry.
- Sanitation: Keep the area clean and free of debris where pests can nest or hide.
- Natural Deterrents: Some advocate for the use of natural deterrents such as peppermint oil or predator urine to ward off rodents.

Natural underground storage techniques offer a sustainable and energy-efficient way to preserve food, tapping into the earth's innate ability to maintain stable conditions conducive to long-term storage. By understanding and applying these age-old methods, we can reduce reliance on electrical preservation methods, decrease food waste, and enjoy seasonal produce well beyond its harvest. Whether through constructing a dedicated root cellar, employing simpler clamp storage, or utilizing earth pits, these techniques connect us to our ancestral roots and the rhythms of the natural world, fostering a deeper appreciation for the bounty it provides and our capacity to sustainably harness it for nourishment and survival.

Book 10:
Living Off the Grid

Chapter 1: Fundamentals of Sustainable Off-Grid Living

Embarking on a journey to live off the grid is a profound declaration of independence and a commitment to sustainability. It's a lifestyle choice that demands a deep connection with the environment, a willingness to embrace simplicity, and a dedication to self-sufficiency. This chapter lays the foundation for sustainable off-grid living, covering the essential principles, planning considerations, and the mindset required to thrive away from conventional utility systems.

Understanding Off-Grid Living

Living off the grid means residing without reliance on public utilities, such as electricity, water supply, and sewage systems. It's about creating a self-sustaining home that meets all your needs without tapping into the grid. This lifestyle not only reduces your carbon footprint but also fosters resilience, independence, and a closer relationship with nature.

The Pillars of Off-Grid Living

Energy Independence

Energy generation and storage are at the heart of off-grid living. Harnessing renewable energy sources—such as solar, wind, and hydro power—enables you to power your home sustainably. Investing in a robust energy storage system, like batteries, ensures you have electricity during periods of low generation.

Water Self-Sufficiency

Securing a reliable water source is critical. Rainwater harvesting, wells, and nearby streams can provide potable water, but it's essential to implement filtration and purification systems to ensure water safety. Sustainable water use practices, such as greywater recycling and drip irrigation, further support water conservation.

Food Autonomy

Growing your own food through sustainable agriculture practices not only guarantees a fresh supply of fruits and vegetables but also reduces dependence on commercial food systems. Techniques like permaculture, organic gardening, and aquaponics can maximize yields while minimizing environmental impact.

Waste Management

Effective waste management is crucial in off-grid living. Composting organic waste, recycling materials, and minimizing consumption reduce landfill contributions and turn waste into resources. Innovative solutions, such as composting toilets, address sanitation needs sustainably.

Planning Your Off-Grid Transition

Assessing Your Needs

Evaluate your energy, water, and food requirements based on your household size, lifestyle, and climate conditions. This assessment guides the design of your off-grid systems and helps determine the resources you'll need to sustain your household.

Location Considerations

Choosing the right location is vital. Factors such as climate, land availability, water access, and local regulations significantly influence your ability to live off the grid. Look for locations that support renewable energy generation and offer fertile soil for gardening.

Building a Sustainable Home

Constructing a home that's efficient, durable, and harmonious with the environment is key. Energy-efficient designs, natural building materials, and technologies like solar panels and rainwater catchment systems are integral to creating a sustainable off-grid home.

Financial Planning

Transitioning to off-grid living requires initial investment in land, infrastructure, and equipment. Budgeting for these expenses, along with ongoing costs for maintenance and living, ensures financial sustainability. Explore incentives and grants available for renewable energy installations and sustainable farming practices.

Embracing the Off-Grid Mindset

Living off the grid demands more than just technical solutions; it requires a shift in mindset.

Adaptability

Off-grid living involves learning and adapting. You'll need to become comfortable with troubleshooting system failures, preserving food, and adapting to the rhythms of nature.

Minimalism

Embracing a minimalist lifestyle reduces consumption and waste, aligning with the principles of sustainability. It's about prioritizing needs over wants and finding value in experiences rather than possessions.

Community Engagement

While off-grid living emphasizes independence, being part of a community is invaluable. Sharing knowledge, skills, and resources with neighbors can enhance your resilience and enrich your off-grid experience.

Continuous Learning

The landscape of sustainable technologies and practices is ever-evolving. Staying informed about advancements in renewable energy, sustainable agriculture, and eco-friendly building techniques allows you to continuously improve your off-grid lifestyle.

The journey to off-grid living is both challenging and rewarding. It offers an opportunity to live in harmony with the environment, reduce your ecological footprint, and cultivate a sustainable lifestyle. By understanding the fundamentals of off-grid living and committing to continuous learning and adaptability, you can build a resilient, self-sufficient home that aligns with your values and aspirations. This chapter has laid the groundwork, providing the essential principles and considerations to guide you on your path to sustainable off-grid living. As we delve deeper into the specifics of harnessing renewable energy, growing your own food, and managing waste sustainably, remember that off-grid living is not just a series of technical tasks but a holistic approach to life that celebrates independence, sustainability, and a deep connection with the natural world.

Chapter 2: Harnessing Renewable Energy in the Wilderness

Living off the grid in the wilderness requires not just a commitment to a sustainable lifestyle but also a profound understanding of how to harness the earth's renewable resources for energy. This chapter explores the various methods and technologies available for generating and storing energy in remote locations, away from conventional power grids. It delves into the practical aspects of utilizing solar, wind, hydro, and biomass energy sources, providing a comprehensive guide to achieving energy independence in the wilderness.

Solar Power: The Sun as a Source of Energy

Solar energy, abundant and accessible in most locations, is a cornerstone of off-grid renewable energy systems.

Photovoltaic (PV) Systems

- Components: PV systems consist of solar panels, a charge controller, batteries for energy storage, and an inverter to convert direct current (DC) to alternating current (AC).
- Installation: Optimal placement is crucial; panels should face true south (in the Northern Hemisphere) and be tilted at an angle equal to the latitude of the location to maximize sun exposure.
- Considerations: Solar energy production is dependent on weather and the seasons, requiring efficient energy storage solutions and possibly a supplementary energy source for times of low sunlight.

Wind Energy: Harnessing the Power of the Wind

Wind energy, with its capability to generate power both day and night, serves as an excellent complement to solar power.

Wind Turbines

- Setup: Wind turbines convert the kinetic energy of wind into electrical power. They require an open, unobstructed area to operate efficiently, with higher placement (often on a tower) capturing stronger winds.
- System Components: Similar to solar, wind energy systems include the turbine, a charge controller, batteries for storage, and an inverter.
- Challenges: Wind availability can be unpredictable, and not all locations are suitable for wind turbines due to low wind speeds or environmental restrictions.

Hydro Power: Utilizing Water Flow

For those situated near a flowing water source, micro-hydro power can offer a consistent and reliable energy solution.

Micro-Hydro Systems

- Mechanism: Water flow turns a turbine, generating electricity. The power output is determined by the flow rate and the vertical drop (head) of the water.
- Installation: Requires access to a stream or river with sufficient year-round flow and head. Systems can range from simple setups with a turbine placed in the stream to more complex installations with diversion channels.
- Advantages: Micro-hydro power can produce electricity continuously, making it one of the most reliable off-grid renewable energy sources.

Biomass Energy: Organic Materials as Fuel

Biomass energy involves using organic materials, such as wood, agricultural crops, or waste, to produce heat or electricity.

Wood Stoves and Biomass Boilers

- Wood Stoves: Provide heating and cooking capabilities using wood as fuel. Modern, efficient stoves ensure complete combustion, reducing smoke and maximizing heat output.
- Biomass Boilers: Can heat water for domestic use or space heating. Some systems can also generate electricity through a process called cogeneration.
- Sustainability: When using wood or other biomass for energy, it's crucial to manage resources sustainably to prevent deforestation and ensure a continuous fuel supply.

Energy Storage and Management

Efficient energy storage and management are critical for maintaining a reliable power supply in an off-grid setup.

Batteries

- Technology: Deep-cycle batteries, such as lead-acid, lithium-ion, or saltwater, store electricity generated by renewable energy sources.
- System Design: Battery capacity should be sized according to your energy needs and the production capacity of your renewable energy systems to ensure adequate power during periods of low generation.

Energy Efficiency

- Conservation: Reducing energy consumption through efficient appliances, LED lighting, and mindful energy use extends the stored energy's lifespan.
- Monitoring: Implementing energy monitoring systems can help track consumption patterns and optimize energy usage, ensuring that your renewable energy system meets your needs effectively.

Harnessing renewable energy in the wilderness is a practical and fulfilling way to achieve off-grid living, providing not just independence from utility grids but also aligning with principles of environmental stewardship and sustainability. By understanding the options available and carefully planning your renewable energy system based on the resources at your location, you can create a reliable, efficient, and sustainable energy solution tailored to your off-grid lifestyle. This chapter has laid the groundwork for exploring renewable energy sources and their application in the wilderness, empowering you to take the next steps toward energy independence and a sustainable future.

Chapter 3: Growing Your Own Food with Sustainable Agriculture

Living off the grid is synonymous with the pursuit of sustainability, and at the heart of this endeavor lies the ability to grow your own food. Sustainable agriculture in an off-grid setting isn't just about self-sufficiency; it's about creating a harmonious relationship with the land, one that nurtures both the soil and the soul. This chapter explores the essentials of cultivating food through sustainable practices, ensuring that your off-grid homestead can thrive with minimal impact on the environment.

Understanding Sustainable Agriculture

Sustainable agriculture is the practice of farming that respects the natural biological processes and cycles of the earth. It aims to produce food in ways that are healthy for consumers and animals, do not harm the environment, provide fair treatment to workers, and support and enhance rural communities.

Principles of Sustainable Agriculture

- Soil Health: Healthy soil is the foundation of sustainable agriculture. Practices such as crop rotation, cover cropping, and the use of organic fertilizers enhance soil fertility and structure, promoting vibrant plant growth.
- Water Conservation: Efficient water use and the conservation of water resources are critical, especially in off-grid settings where access to water may be limited. Techniques like drip irrigation and rainwater harvesting can significantly reduce water usage.
- Biodiversity: Encouraging a diverse ecosystem around and within the crop areas supports natural pest control, pollination, and disease prevention, reducing the need for chemical inputs.

Planning Your Sustainable Garden

Assessing Your Land

Begin by understanding your land's capabilities and limitations. Assess soil quality, sunlight exposure, water availability, and wind patterns. This initial assessment will guide your garden design, crop selection, and sustainability strategies.

Garden Design

Design your garden to work with the natural characteristics of your land. Consider raised beds to improve drainage and soil quality, companion planting to enhance growth and deter pests, and permaculture principles to mimic natural ecosystems.

Soil Management

Healthy soil is teeming with life and rich in organic matter. Build your soil's health by:

- Composting: Turn kitchen scraps, yard waste, and animal manures into nutrient-rich compost to feed your soil.
- Mulching: Use organic mulches to conserve moisture, suppress weeds, and add organic matter to the soil as they decompose.
- No-till Gardening: Minimize soil disturbance to preserve soil structure, protect soil life, and reduce erosion.

Water Management

In an off-grid lifestyle, managing water wisely is paramount. Employ strategies that maximize efficiency and minimize waste:

- Rainwater Harvesting: Collect rainwater from rooftops for garden irrigation.
- Drip Irrigation: Deliver water directly to the plant roots, reducing evaporation and runoff.
- Swales and Contour Planting: Use landscape features to capture and distribute rainwater effectively.

Crop Selection and Rotation

Choose crops that are well-suited to your climate and soil conditions. Diversity in crop selection enhances nutritional variety and supports ecological balance. Practice crop rotation to prevent soil depletion and disrupt pest and disease cycles.

Integrated Pest Management (IPM)

IPM emphasizes the use of natural pest control methods:

- Beneficial Insects: Encourage or introduce insects that prey on harmful pests.
- Physical Barriers: Use floating row covers or insect netting to protect plants.

- Natural Pesticides: When necessary, opt for organic or homemade pesticides that are less harmful to the environment.

Harvesting and Preserving

Maximize the bounty of your garden by harvesting at peak ripeness for the best nutritional value and flavor. Preserve excess produce through canning, drying, fermenting, or freezing to enjoy the fruits of your labor throughout the year.

The Role of Livestock

Integrating livestock into your off-grid homestead can enhance sustainable agriculture efforts:

- Manure: Animal waste is a valuable source of organic fertilizer.
- Grazing: Properly managed grazing can improve soil health and reduce the need for mowing and weeding.
- Integrated Systems: Animals can contribute to a closed-loop system, reducing waste and increasing efficiency.

Challenges and Solutions

Sustainable off-grid farming is not without its challenges, such as labor intensity, pest and disease management, and climate variability. Solutions include leveraging community resources, continuous learning and adaptation, and integrating technology where appropriate, such as solar-powered irrigation systems.

Growing your own food through sustainable agriculture is a cornerstone of living off the grid. It demands not only a commitment to environmental stewardship but also a willingness to engage deeply with the land. By adopting practices that promote soil health, water conservation, biodiversity, and ethical animal husbandry, you can cultivate a garden that nourishes both body and earth. This chapter provides the groundwork for establishing a sustainable food system on your off-grid homestead, one that ensures resilience, abundance, and a lasting connection to the natural world.

Chapter 4: Effective Waste Management and Recycling Practices

Embracing an off-grid lifestyle involves not just a departure from conventional utility services but also a fundamental rethinking of waste management. In the absence of municipal waste services, effective waste management and recycling practices become essential components of sustainable living. This chapter explores strategies for reducing, reusing, recycling, and responsibly disposing of waste in an off-grid setting, ensuring that your footprint on the natural world is as light as possible.

Understanding Waste in Off-Grid Living

Living off the grid confronts you with the direct impact of your consumption habits. Every item brought into this lifestyle must be accounted for, from its utility and longevity to its eventual disposal or repurposing. Effective waste management in an off-grid context emphasizes the minimization of waste production and the innovative reuse and recycling of materials.

The Principles of Off-Grid Waste Management

- Reduce: The first and most effective step in waste management is to reduce the amount of waste generated. This involves mindful purchasing, prioritizing quality and durability over convenience, and avoiding single-use products.
- Reuse: Before disposing of an item, consider whether it can be repurposed. Reusing materials not only reduces waste but also can inspire creativity in finding new uses for old items.
- Recycle: For waste that cannot be avoided or reused, recycling becomes the next best option. While off-grid living may limit access to municipal recycling facilities, many materials can be recycled or upcycled on-site.
- Responsibly Dispose: Finally, for waste that cannot be reduced, reused, or recycled, responsible disposal methods must be considered to minimize environmental impact.

Reducing Waste Production

- Composting Organic Waste: Kitchen scraps, garden waste, and even certain types of paper can be composted to create nutrient-rich soil for gardening, turning potential waste into a valuable resource.
- Buying in Bulk: Purchasing food and other consumables in bulk reduces packaging waste and often saves money in the long run.

- Choosing Sustainable Products: Opt for products made from sustainable, biodegradable materials whenever possible.

Reusing and Repurposing

- Creative Repurposing: Old containers can become planters, scraps of fabric can be turned into quilts or cleaning rags, and pallets can be transformed into furniture.
- Maintenance and Repair: Learning basic repair skills can significantly extend the life of tools, appliances, and clothing, reducing the need to consume.

Recycling On-Site

- Glass and Metals: While more challenging to recycle at home, glass and metals can often be repurposed. Glass jars become storage containers, and metal cans can be used for a variety of DIY projects.
- Plastics: While reducing plastic use is ideal, some off-grid homesteads find innovative ways to reuse plastic containers. In some cases, specialized machines can melt and reform plastic into new items.
- Paper and Cardboard: Beyond composting, paper and cardboard can be reused as mulch in the garden or as material for paper mache crafts.

Responsible Disposal

- Human Waste: Composting toilets offer a sustainable solution for sewage management, transforming waste into compost that can be used to fertilize non-edible plants.
- Non-Compostable and Non-Recyclable Waste: For waste that cannot be composted or recycled, research the most environmentally friendly disposal methods available in your area. This may involve occasional trips to recycling centers or waste disposal facilities.

Waste Reduction Strategies

- DIY and Homemade Products: Making your own cleaning products, toiletries, and food items from raw ingredients reduces packaging waste and gives you control over the materials used.
- Digital Solutions: Opt for digital subscriptions, billing, and communication where possible to reduce paper waste.

Community Engagement

- Sharing Resources: Collaborate with neighbors and local communities to share resources, tools, and skills, reducing the need for individual consumption.
- Community Recycling Programs: Participate in or initiate community recycling programs for materials that require professional processing, such as electronics or hazardous waste.

Effective waste management and recycling practices are crucial for maintaining the sustainability of off-grid living. By adopting a conscientious approach to consumption, prioritizing the reduction of waste, and finding innovative ways to reuse and recycle materials, off-grid residents can significantly lessen their environmental impact. This chapter has provided a comprehensive guide to managing waste in a responsible, efficient, and environmentally friendly manner, ensuring that living off the grid can be a truly sustainable and fulfilling endeavor.

Made in the USA
Middletown, DE
22 March 2024